Ecosystem Chaos: Invasive Species Impact

Molina

Table of Contents

GENERAL INTRODUCTION

The introduction of non-indigenous species into foreign habitats where they can establish and spread is a major ecological hazard (Keller et al., 2011, Lowry et al., 2013, Simberloff et al., 2013, Mollot et al., 2017). Introductions are promoted or facilitated by anthropogenic long-distance transport of live specimens. The primary reasons for intentional translocations are livestock and pet trade and sport activities, such as angling and hunting, although species may also be unintentionally translocated as stowaways or contaminants, e.g. in ballast water (Padilla & Williams, 2004, Hulme et al., 2008, Keller et al., 2009, Keller et al., 2011). A large number of fish species, such as tilapia (*Oreochromis niloticus*), the European perch (*Perca fluviatilis*), or the common carp (*Cyprinus carpio*) were either translocated widely for aquaculture or ornamental trade from which they escaped into the wild or they were intentionally introduced into novel environments for angling. Once established in the natural environment, they were shown to outcompete native species for resources (e.g. food or space), to prey on native species, and to interfere with species interactions, thereby threatening the native fish fauna (Zambrano et al., 2001, Lintermans, 2004, Britton & Orsi, 2012). Similarly, several American and Australian crayfish species were traded for aquaculture and as ornamental aquarium and garden pond species in Europe and escaped into natural freshwater habitats. They were also purposely stocked to support crayfish harvest. These aggressive and highly competitive non-indigenous crayfish now pose a serious risk to the native crayfish diversity (Chucholl, 2013, Holdich et al., 2009). They also propagate the infectious crayfish plague (see below; Svoboda et al., 2017). The red-eared slider (*Trachemys scripta elegans*) is a popular ornamental turtle species that was released into the wild. In Europe, it is now a strong competitor of endangered native turtle species for food and optimal habitats (Cadi & Joly, 2004, Polo-Cavia et al., 2010). Squirrels were translocated across the world for hunting and fur production and several species have become invasive and are now threatening the persistence of native squirrels through competition and spreading disease (Bertolino, 2009, Bertolino et al., 2014). A huge number of species such as the zebra mussel (*Dreissena polymorpha*), the golden mussel (*Limnoperna fortunei*), the Chinese mitten crab (*Eriocheir sinensis*), or the diatom *Didymosphenia geminata* were accidentally transported to new geographic regions and are now of ecological and economic concern because they are transforming ecosystems and destroying hard structures (Karatayev et al., 2007, Dittel & Epifanio, 2009, Reid et al., 2012).

One group of organisms that has been given little consideration when it comes to invasive species are parasites. Parasites live in or on their host for some or all of their life stages, feed off the host, and reduce host fitness (Zelmer, 1998). Parasites, broadly defined to include diverse groups such as viruses, bacteria, fungi, protozoans, metazoans, and plants, are mostly introduced accidentally with infected hosts due to lack of disease control measures for translocated animals and plants and their products (Cunningham, 1996, Fèvre et al., 2006, Kock et al., 2010, Tompkins et al., 2015). For parasites to become invasive, they need to establish self-sustaining populations by exploiting suitable host species in the new environment. This often involves switching to novel host species in which they can survive and reproduce (Lymbery et al., 2014). Parasite translocation and acquisition of novel hosts are important drivers for the dramatic increase in disease emergence in the last decades (Daszak et al., 2000, Peeler et al., 2011). Additionally, ecosystem degradation, e.g. through overexploitation, contamination, or fragmentation, can facilitate the establishment of invasive parasites (Daszak et al., 2001, Okamura & Feist, 2011).

A number of translocated parasites have been reported for terrestrial species, mostly associated with translocation of livestock, pets, and for conservation (Cunningham, 1996, Fèvre et al., 2006, Kock et al., 2010). Rabies had been introduced into a raccoon population in Virginia, USA, via translocation of rabid individuals from Florida for restocking. Rabies had also been introduced into a previously rabies-free Indonesian island by rabid dogs. In both cases it led to epidemics and is now endemic and a health threat to wildlife and humans (Jenkins et al., 1988, Fèvre et al., 2006). Rinderpest was introduced into Europe and Africa from Asia for centuries by cattle accompanying armies in their invasions and later by cattle trade. The rinderpest epidemic in sub-Saharan Africa at the end of the 19th century caused high mortality in wild ungulates, livestock, and humans (Roeder et al., 2013). Bat colonies in North America have suffered from high mortalities after the introduction of the fungus *Pseudogymnoascus destructans* causing white-nose syndrome (Frick et al., 2010) and in Hawaii lowland avifauna collapsed following the introduction of avian malaria at the beginning of the 20th century and has not yet recovered (van Riper et al., 1986, Samuel et al., 2015). The spill-over and subsequent spread of the mite *Varroa destructor* from the Asian honey bee (*Apis cerana*) to the European honey bee (*A. mellifera*) is associated with the mass mortalities of honey bees worldwide (Le Conte et al., 2010).

There are also many known examples of parasites successfully invading freshwater habitats and infecting novel hosts. The vast majority of aquatic emerging diseases are associated with translocations of infected hosts for aquaculture and ornamental fish trade and subsequent spill-over of parasites into adjacent habitats (Peeler & Feist, 2011, Adlard et al., 2015, Tompkins et al., 2015). Their establishment may be facilitated by the long history of large-scale degradation and disturbance of freshwater systems by humans (Okamura & Feist, 2011). Hence, a larger number of emerging vertebrate parasites has been reported for fish than for any other family (Tompkins et al., 2015). The Asian fish tapeworm *Schyzocotyle acheilognathi*, for example, has been introduced worldwide by fish trade for aquaculture and it has acquired more than 300 fish species as new hosts, making it one of the most successful invasive parasite species. In aquaculture, it was reported to cause mortality of infected fish (Kuchta et al., 2018). The generalist fish parasite *Sphaerothecum destruens* was introduced into Europe, possibly with salmonids for aquaculture from North America. In Europe, it causes high mortality in various fish species (Gozlan et al., 2005, Ercan et al., 2015). The introduction of *Myxobolus cerebralis* from Europe into North America with infected salmonids in the mid-1950s has decimated the salmonid populations in several North American rivers (Bergersen & Anderson, 1997, Hedrick et al., 1998).

Diseases of aquatic animals other than fish have also been reported to cause devastating population declines. Chytridiomycosis, caused by the fungus *Batrachochytrium dendrobatidis*, one of the best known emerging diseases of wildlife, is decimating amphibian populations worldwide and has already lead to species extinctions. Amphibian trade facilitates the spread of the disease (Lips et al., 2006, Skerratt et al., 2007, McKenzie & Peterson, 2012, Scheele et al., 2019). The crayfish plague (*Aphanomyces astaci*) introduced into Europe with American crayfish, is responsible for mass mortalities and local extinction of native crayfish species (Alderman et al., 1987, Kozubikova et al., 2007, Svoboda et al., 2017, Vralstad et al., 2014). The spillover of the blood fluke *Spirorchis elegans* from the red-eared slider to a native European turtle (*Emys orbicularis*) was associated with a mass mortality event in Galicia, Spain (Iglesias et al., 2015).

Aside from causing direct mortality, invasive parasites can disturb host physiology and behaviour, diminish genetic diversity and alter evolutionary trajectories, and render hosts more susceptible to secondary stressors (Shirakashi et al., 2008, Kleindorfer & Dudaniec, 2016, McKnight et al., 2017). The absence of a shared evolutionary history between novel hosts and parasites is proposed as one reason for the negative health, fitness effects, and failing response of novel hosts (Taraschewski, 2006, Mastitsky et al., 2010).

Host-parasite interaction

Parasitism is the most common lifestyle and there is at least one parasite in any known species (Windsor, 1998, Dobson et al., 2008). During the long-standing relationship between hosts and parasites, hosts evolve defence mechanisms to fend off the parasites, which in turn evolve mechanisms to evade the host response. This leads to continuous co-evolution between hosts and parasites in which both try to maximise their fitness (Anderson & May, 1982, Woolhouse et al., 2002). Parasites can indeed impose strong selective pressures that drive rapid host adaptation as indicated by frequency shifts of major histocompatibility complex (MHC) class IIB alleles towards the resistance alleles only one generation after experimental infections of three-spined sticklebacks (*Gasterosteus aculeatus*) (Eizaguirre et al., 2012b). Similarly, the frequency of a salmon (*Salmo salar*) MHC IIB allele associated with a higher infection rate by a myxozoan parasite decreased markedly during the summer season in the St. Lawrence Estuary where this parasite was the most prevalent (Dionne et al., 2009). Studies on temporal changes of the genetic composition of invertebrate host populations also support parasite-mediated selection, both in the lab and in the wild. In clonal snail (*Potamopyrgus antipodarum*) populations and in *Daphnia* populations the most common clone declined over time in infected populations which was not seen in uninfected populations (Duncan & Little, 2007, Koskella & Lively, 2009, Wolinska & Spaak, 2009).

A common evolutionary history appears to be important for coping with a particular parasite or parasite community. Marine three-spined sticklebacks, the ancestors of freshwater sticklebacks, suffer from higher parasite loads when transferred to freshwater environments than their freshwater conspecifics (MacColl & Chapman, 2010) and experimental infections of marine sticklebacks with a freshwater cestode parasite (*Schistocephalus solidus*) indicated that they are highly susceptible (Weber et al., 2017). Marine populations have never been in contact with freshwater parasites and could therefore not acquire resistance. Furthermore, a higher probability of encountering parasites may increase resistance. A German three-spined stickleback population, where natural prevalence with *Schistocephalus solidus* is low, can get more highly infected than a Norwegian population, where prevalence is higher (Kalbe et al., 2016). River populations are consistently more resistant to *Gyrodactylus* sp., predominantly riverine parasites, than lake populations. This was linked to their MHC IIB genetic composition (Eizaguirre et al., 2011, Eizaguirre et al., 2012a). Also, invertebrate populations that had experienced epidemics were more resistant to reinfections than those without recent epidemics (Schoebel et al., 2010, Duffy & Sivars-Becker, 2007).

Host defence response

Hosts have developed several mechanisms to defend themselves against parasites. Hosts can reduce the risk of becoming infected through behavioural avoidance. This involves avoiding sites of parasite aggregation, contaminated food, or infected conspecifics but also avoiding the establishment of parasite infections by grooming, cleaning, or preening (Behringer et al., 2018, Sarabian et al., 2018).

Following infection, hosts can reduce or clear infections by activating the immune system, this is termed resistance (Schneider & Ayres, 2008, Raberg et al., 2009). In vertebrates, the immune system consists of two interdependent branches, the innate and the adaptive immune system (Fig. 1; Murphy & Weaver, 2017, Abbas et al., 2018). The innate immune system is fast and relatively unspecific, responding to a wide variety of parasites. These are recognized by omnipresent leukocytes via receptor-binding to patterns that are shared among different groups of parasites. Upon recognition, the leukocytes produce inflammatory mediators, such as cytokines and chemokines. This initiates an inflammatory response that recruits further leukocytes to the site of infection. Leukocytes ingest and destroy invading parasites via phagocytosis and destroy them within specialized vesicles called lysosomes. Parasites too large to be phagocytosed are destroyed by the release of antimicrobial substances. The complement system forms a membrane attack complex which disrupts the cell membrane of the parasite, leads to its lysis, and enhances phagocytosis by leukocytes (Murphy & Weaver, 2017).

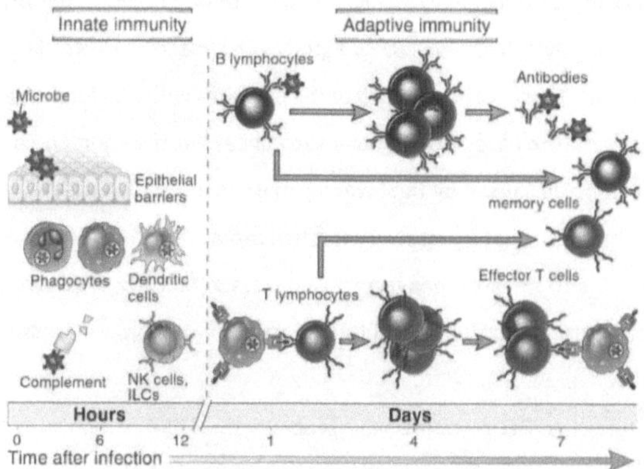

Figure 1 Components of the innate and adaptive immune system. The innate immune mechanisms respond within minutes to hours of an infection. The adaptive immune mechanisms require several days to fully develop. Modified from Abbas et al. (2015).

Dendritic cells, and to a lesser extent macrophages and B lymphocytes, recognize and tag elements of an invading parasite, called antigens, to initiate the response of the adaptive immune system. These cells migrate to secondary lymphoid organs and present antigens bound to MHC molecules on their surface. Within the secondary lymphoid organs, antigen-MHC complexes are recognized by T cell receptors (TCR) of naïve T lymphocytes. Following activation, T lymphocytes proliferate to produce identical clones, differentiate into effector cells, and migrate to the site of infection (Murphy & Weaver, 2017). The effector function of these cells depends on the type of the invading parasite. Parasites entering the cytoplasm of host cells (intracellular, mostly viruses) typically cause differentiation into cytotoxic T cells which induce apoptosis of infected cells (Parkin & Cohen, 2001). Extracellular parasites stimulate differentiation into helper T cells. These provide the signals necessary for B lymphocytes that have antigens bound to their B cell receptor (BCR) to proliferate and differentiate into antibody (Ab)-secreting cells (Parkin & Cohen, 2001, Crotty, 2015). The adaptive immune response is highly specific to the invading parasite due to the huge repertoire of TCRs and BCRs but it is slow in responding, taking about 1 week to be fully mounted. However, it leads to the formation of immunological memory through the development of long-lived memory T and B cells. Upon re-encountering the same parasite, these memory cells can induce an immediate adaptive immune response that is typically stronger than the first (Murphy & Weaver, 2017, Abbas et al., 2018).

Possibly the ability to form an immunological memory is partially responsible for higher resistance in (vertebrate) populations that have experienced the infecting parasite before.

Distinct types of helper T cells respond to different types of extracellular parasites and determine the class of Abs that are produced. The type of helper T cell further determines the type of feedback on the innate immune system. Intravesicular parasites (e.g. mycobacteria) induce the production of type 1 helper T cells (Th1). These stimulate macrophages to increase their lysosomal activity and clear the infection. Furthermore, Th1 cells recruit additional macrophages to the site of infection, thereby increasing the inflammatory response (Murphy & Weaver, 2017). Extracellular macroparasites (e.g. helminths) induce the production of Th2 cells. These induce coating of the parasites with Abs for better recognition and destruction by leukocytes that release antimicrobial substances. Furthermore, a Th2 response stimulates the repair of tissue damage caused by the invading macroparasite (Díaz & Allen, 2007, Gause et al., 2013). Extracellular microparasites (e.g. bacteria or fungi) promote a Th17 response. Th17 cells stimulate Ab-coating of parasites to enhance phagocytosis and lysis (Murphy & Weaver, 2017). Th1, Th2, and Th17 responses are mutually exclusive and the induction of one type represses the induction of the others (Murphy & Weaver, 2017).

Due to their pro-inflammatory nature, the Th1 response, and probably the Th17 response, are highly destructive. This is required to limit microparasitic replication, which is usually fast and occurs within the host (Graham et al., 2005, Murphy & Weaver, 2017). However, inflammatory mediators do not discriminate between host cells and parasitic cells and cause considerable damage to the host (i.e. immunopathology; Graham et al., 2005). A Th1 response is capable of clearing macroparasitic infections but due to their large size, eliminating them would cause excessive damage to the host as well (Graham et al. 2005). Macroparasites often do not replicate within the host and if so, at a much lower rate than microparasites. Thus, fast clearing of the macroparasite infection is often not necessary for survival. The need to minimize and repair damage caused by both immunopathology and parasites may have led to the evolution of the anti-inflammatory Th2 response (Graham et al., 2005).

Besides resisting parasites, hosts can tolerate infections by minimizing negative fitness effects of harbouring an infection (Raberg et al., 2009, Medzhitov et al., 2012). More tolerant hosts will thus have higher fitness for a given parasite load than less tolerant hosts. Contrary to avoidance and resistance, tolerance does not reduce parasite fitness, thus it should not lead to antagonistic coevolution. Resistance and tolerance are not mutually exclusive and together determine host fitness once an infection has established. However, evidence suggests that higher tolerance may be associated with lower resistance and vice versa (Raberg et al., 2007, Klemme & Karvonen, 2017).

All three major defence mechanisms are associated with fitness costs (Sheldon & Verhulst, 1996, Raberg et al., 2009, Sadd & Schmid-Hempel, 2009, Behringer et al., 2018). This may explain why variability in those traits is maintained within and among species and populations and why the strongest (immune) response is not necessarily the best. The optimal response can depend on the abiotic environment, such as temperature or food availability, and on the biotic environment, including predation risk or variable parasite communities, and all three mechanisms have been shown to vary among populations from different environments (Sadd & Schmid-Hempel, 2009, Sorci, 2013, Klemme & Karvonen, 2017).

Gene expression in the host response

Gene expression involves several steps: first, polymerase- and transcription factor-binding initiates transcription from DNA that produces an RNA copy of the encoding gene. The RNA is then spliced into mRNA and subjected to post-transcriptional modifications. Then, the mRNA is translated into an amino acid chain which folds into a protein and finally, the protein undergoes post-translational modifications. Regulation of these steps determines the amount and efficiency of the protein (Plotkin, 2010). Modification of gene expression is fast and enables plastic responses when facing changing conditions. Gene expression, although denoting the production of the functional protein in its strict sense, is more commonly used to refer to the production of mRNA and this is how it will be used throughout the rest of the book.

Gene expression has been widely applied to the study of host-parasite interactions. Both transcriptome-wide studies and studies restricted to pre-selected candidate genes revealed that parasite infections readily lead to expression changes. In controlled experimental infections, numerous immune processes are affected and the transition from innate to adaptive immune processes over time was observed in mice infected with *Eimeria falciformis* (Ehret et al., 2017). Also, non-immune processes can be strongly modified in infected individuals. Experimental infections of host populations or genotypes that differ in their susceptibility provide evidence that this pattern can at least partly be explained by differences in gene expression (Lenz et al., 2013, Guo et al., 2016, Lohman et al., 2017) and this can also be seen in the natural environment between populations that vary in susceptibility (Huang et al., 2016). However, expression of immune genes can be considerably plastic and rapidly adapt to a new environment (Stutz et al., 2015).

Host response to invasive parasites

The regulation of an immune response may be a core determinant of the higher susceptibility of novel host species and an incompletely or inappropriately activated immune system may be damaging to the host rather than protective. A North American bat species affected by white-nose syndrome has been found to increase expression of inflammatory genes locally. However gene expression of the adaptive immune system cannot be induced during hibernation (Field et al., 2015). This can lead to an exaggerated immune response following emergence from hibernation that may then cause severe wing lesions and death (Meteyer et al., 2012). In frogs susceptible to chytridiomycosis, transcriptome-wide gene expression suggests that the immune system is not consistently activated (Rosenblum et al., 2012, Ellison et al., 2015) and that the activation does not reduce parasite load and confer protection (Poorten & Rosenblum, 2016, Eskew et al., 2018). In European honey bees, gene expression indicates suppression of an immune response upon infection with the microsporidian *Nosema ceranae* (Aufauvre et al., 2014). On the other hand, infections with *Varroa* mites do not appear to result in differential immune gene expression in European honey bees but they do in Asian honey bees (Zhang et al., 2010). In European honey bees, *Varroa* mite infections result in differential expression of a variety of metabolic genes (Zhang et al., 2010). Infected susceptible frogs and bats also modified processes related to metabolism and to energy generation and consumption (Field et al., 2015, Poorten & Rosenblum, 2016, Eskew et al., 2018). In contrast, resistant species modulate expression of immune and other genes to a far lesser extent (Ellison et al., 2015, Poorten & Rosenblum, 2016, Davy et al., 2017, Eskew et al., 2018).

The severe impact of invasive parasites may promote adaptation by novel hosts on an ecological timescale (Penczykowski et al., 2011). There are indeed several examples where resistance or tolerance has evolved within a few decades. Rainbow trout (*Oncorhynchus mykiss*) infected with the myxozoan parasite *Myxobolus cerebralis* evolved increased resistance within a decade of the parasite's introduction (Miller & Vincent, 2008). Twelve years after a spill-over event of the bacterium *Mycoplasma gallisepticum* to a house finch (*Carpodacus mexicanus*) population, finches were more resistant than a population that had never been exposed. Resistance appeared to be associated with an immune response since immune gene expression was up-regulated in the more resistant population (Bonneaud et al., 2011, Bonneaud et al., 2012). The honeycreeper species Hawai'i 'Amakihi (*Hemignathus virens*) did not increase in resistance to avian malaria (*Plasmodium relictum*) but rather in tolerance. This allowed the population to recover from drastic reductions that occurred several decades earlier (Atkinson et al., 2013). Populations may evolve along different trajectories despite invasion by the same parasite. Blue mussels (*Mytilus edulis*) infected by an invasive copepod (*Mytilicola intestinalis*) evolved resistance along one invasion front but tolerance along another invasion front spreading from the same point of introduction (Feis et al., 2016). These different trajectories are also partly reflected in differential gene expression of both host and parasite (Feis et al., 2018).

Parasites shape species communities

Despite their negative effect on hosts, parasites are essential players in community composition and ecosystem functioning (Hatcher et al., 2012, Preston et al., 2016). Parasites have the potential to modify the abundance, life history, and behaviour of their hosts which influences competition, food-web structures, and trophic cascades (Kohler & Wiley, 1997, Wood et al., 2007, Lefèvre et al., 2009). Depending on the host's position within a community, this can lead to drastic changes in species composition and interactions.

Aside from structuring communities of free-living species, parasites can influence each other (Poulin, 1999). Within a host individual, parasite communities undergo similar ecological interactions as free-living species within their community. Parasites can compete for resources or engage in predator-prey interactions. These competitions can be both intra- and inter-specific. Through the manipulation of hosts, parasites can indirectly influence parasite communities. A trophically transmitted parasite that is not capable of manipulating host behaviour may hitch-hike with a manipulative parasite sharing the same life cycle, thereby increasing its transmission efficiency (Thomas et al., 1997). On the other hand, parasite manipulation of co-infecting parasites can lead to competition over host behaviour manipulation (Hafer & Milinski, 2016). Both facilitating and conflicting behaviour manipulations can affect parasite abundance. Parasites can also influence each other through the activation of the host immune system which can be beneficial or disadvantageous to a co-infecting parasite (Page et al., 2006). The outcome of co-infections will ultimately determine the extent to which parasites influence community composition of free-living species and ecosystem functioning across time and space. The impact of invasive parasites on novel hosts, and possibly on the prevailing parasite community, can then lead to alterations of ecological communities and ecosystem functioning via cascading effects.

Study system

Temperate freshwater eels are susceptible to a wide range of stressors, including parasites, and this has synergistically led to population declines over the last four decades (Drouineau et al., 2018). Recruitment of the European eel (*Anguilla anguilla*) declined to 2.1-29% of the mean recruitment level from 1960-1979, depending on region and developmental stage, and recovery is not seen so far (ICES, 2018). Recruitment of Japanese eels (*A. japonica*) declined by 80% and that of the American eel (*A. rostrata*) is similarly low (Drouineau et al., 2018). All three species are listed as endangered or critically endangered on the IUCN Red List (IUCN, 2018). Eels are catadromous species. They spend up to 20 years foraging in continental fresh waters before reproducing in the marine environment (Fig. 2;

Schmidt, 1923, Tesch, 2003, Aoyama et al., 2014). Alterations to oceanic currents interfere with larval drift to the continent (Friedland et al., 2007, Baltazar-Soares et al., 2014). River constructions hamper upstream migration of juveniles and downstream migration of silver eels that leave continental waters for spawning (Tesch, 2003, Miller et al., 2016). During the continental stage, they are affected by overfishing (Dekker, 2003), contamination (Geeraerts & Belpaire, 2010), and infection with non-native parasites (Drouineau et al., 2018, Miller et al., 2016).

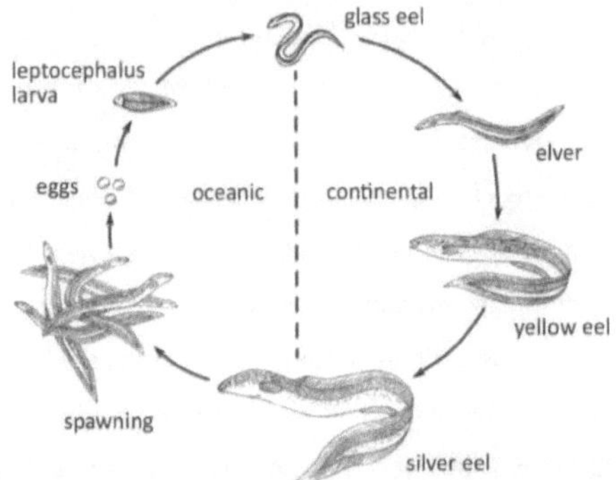

Figure 2 Life cycle of eels. Eels spawn in the ocean. The leptocephalus larvae are drifted to the continental shelf by oceanic currents. Upon reaching continental waters, the larvae metamorphose into glass eels and enter freshwater systems. Glass eels develop into elvers and migrate upstream. The eels then spend approximately one decade foraging as yellow eels before metamorphosing into silver eels and migrating back to the oceanic spawning site.

The arrival of the parasitic swim bladder nematode *Anguillicola crassus* coincides with the steep population decline of the European eel (Kirk, 2003). The parasite is native to the Japanese eel and it was introduced into the European eel population with live imports for aquaculture in the late 1970s and early 1980s. Following importation, it escaped into the wild and spread across the European

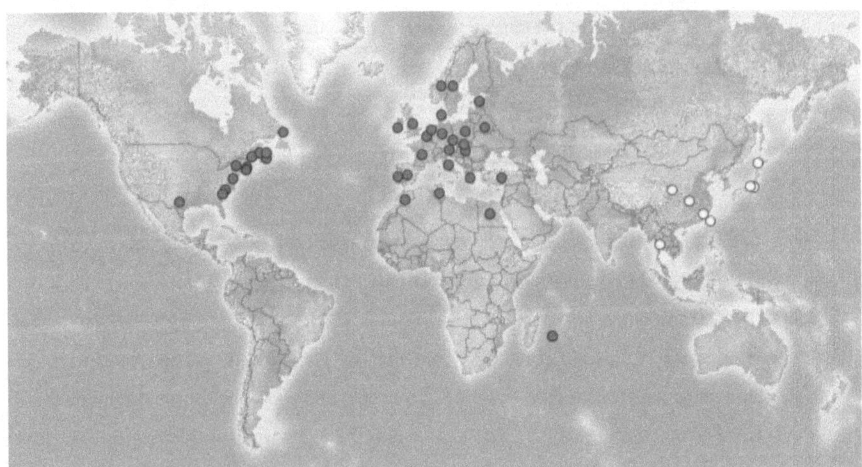

Figure 3 Worldwide distribution of *A. crassus*. Red dots indicate areas where *A. crassus* is invasive, white dots indicate its native range. Map retrieved and modified from the Invasive Species Compendium (https://www.cabi.org/isc).

continent and North Africa within a few years due to eel trafficking for stocking and aquaculture and it is now present in most of the European eel's native range (Fig. 3; Kirk, 2003, Taraschewski, 2006). Thus, the parasite *A. crassus* has existed in the European eel for approx. 3-5 eel generations. In the mid-1990s it was introduced into wild American eels following import of infected eels for aquaculture and it is spreading across the American Atlantic coast (Fig. 3; Aieta & Oliveira, 2009, Hein et al., 2014).

Anguillicola crassus is a trophically transmitted parasite with an obligate intermediate copepod host, a facultative paratenic fish host, and eels as final hosts (Fig. 4; De Charleroy et al., 1990, Nagasawa et al., 1994). It can be acquired by eels from first feeding as glass eels throughout their entire continental phase (Nimeth et al., 2000). Upon release into fresh water, second stage larvae (L_2) hatch and are ingested by copepods. In the copepod, they moult into third stage larvae (L_3) which is the infective stage for eels (Bonneau et al., 1991). Eels get infected by feeding on infected copepods or paratenic hosts. In the latter, *A. crassus* survives and remains infective for eels but it does not develop to adulthood (Thomas & Ollevier, 1992). Following ingestion by eels, *A. crassus* L_3 migrate from the intestine to the swim bladder wall where they can be first observed within a few days (Knopf et al., 1998). After several days to weeks, they moult into fourth stage larvae (L_4; Blanc et al., 1992, Moravec et al., 1994a). They then migrate into the swim bladder lumen where they mature into sexually reproducing adults. The eggs are released into the water with the faeces. Under laboratory conditions, the life cycle can be completed within less than two months (De Charleroy et al., 1990, Weclawski et al., 2013).

Anguillicola crassus reaches higher prevalence (proportion of infected hosts) and infection intensity (number of parasites per infected host) in the European eel (e.g. Audenaert et al., 2003, Lefebvre & Crivelli, 2004, Schabuss et al., 2005, Gérard et al., 2013) than in the Japanese eel (Nagasawa et al., 1994, Münderle et al., 2006, Heitlinger et al., 2009). Shortly after its introduction into Europe, *A.*

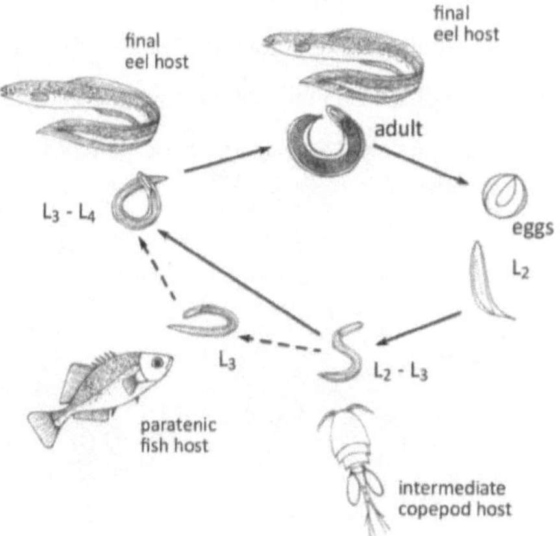

Figure 4 Life cycle of *Anguillicola crassus*. The parasite reproduces sexually in the swim bladder of its final eel host. Eggs are released into the water with the faeces. In the water, the second stage larvae (L_2) hatch. They are ingested by the intermediate copepod host in which they develop into L_3. When eels feed on infected copepods or infected paratenic fish hosts, L_3 are transmitted to the final eel host. L_3 migrate from the intestine to the swim bladder wall, develop into L_4, and finally mature into blood-feeding male and female adults that reside in the swim bladder lumen and reproduce.

crassus was associated with a mass mortality event in a reservoir with poor water condition and high eel density and infection pressure (Molnár et al., 1991). Hypoxia was shown to increase mortality rates of severely infected individuals (Molnár, 1993, Lefebvre et al., 2007). Infection also increases oxygen consumption and affects swimming speed and behaviour of migrating eels (Palstra et al., 2007, Sjöberg et al., 2009, Newbold et al., 2015). This may prevent them from reaching the spawning area and reproduce (Pelster, 2015). In contrast, no adverse effects have been reported on the Japanese eel. Even in potentially stressful aquaculture conditions, they seem to cope well with *A. crassus* infections (Egusa, 1979).

Transcriptome-wide gene expression studies of naturally infected eels at all continental life stages (yellow eels, silvering eels, and silver eels), indicate that a wide variety of processes are modified in the swim bladder after acclimation to laboratory conditions (Pelster et al., 2016, Schneebauer et al., 2017). These include processes involved in metabolism, swim bladder function, and immunity. Local inflammation and immune cell aggregation were also observed in morphological examinations of infected swim bladders (Molnár et al., 1993, Würtz & Taraschewski, 2000) and the presence of Abs in blood and head kidney suggests the induction of a systemic immune response to adult *A. crassus* (Nielsen, 1999, Knopf & Lucius, 2008). Abs are also produced by the Japanese eel (Nielsen, 1999, Knopf & Lucius, 2008); however, Ab concentrations do not correlate with infection intensity (Knopf & Lucius, 2008).

The native Japanese eel host responds to *A. crassus* by encapsulating it (Münderle et al., 2006, Heitlinger et al., 2009). This response is considered to be a major reason for lower infection intensities in the Japanese eel compared to the European eel (Knopf & Mahnke, 2004, Keppel et al., 2014) and prevalences of capsules of up to 22% were reported for wild Japanese eels (Münderle et al., 2006, Heitlinger et al., 2009). In the European eel, *A. crassus* can also get encapsulated, however, there are only occasional reports (Molnár, 1994, Audenaert et al., 2003). Nonetheless, Audenaert et al. (2003) reported an increase of encapsulated *A. crassus* in Flanders, Belgium, from none in 1990 to 20% of all retrieved *A. crassus* in 2000. Despite this being the response observed in the native host, the authors suggested encapsulation in the European eel to be an additional cost which can further impair host health rather than an adaptation to the novel parasite.

OBJECTIVES

The first overall objective of the book was to determine which processes contribute to higher susceptibility of novel hosts compared to native hosts upon parasite infection. The different outcomes of *A. crassus* infections in the native Japanese eel host and the novel European eel host have thus far been attributed to an effective immune response by the native host, but not by the novel host. The mechanisms underlying the responses of the two eel species to *A. crassus* have not been identified so far and the immediate first response upon infection is severely understudied. Thus, one question that this book intended to answer was whether an early immune activation is key to a successful defence response. Evidence from other vertebrate-invasive parasite systems suggests that both immune and non-immune gene expression is induced in susceptible host species, but not resistant host species. A second question for this book was if gene expression of eels in response to *A. crassus* follows a similar pattern.

The second objective was to determine if novel hosts benefit from the evolution of a defence response in the natural environment. Defence mechanisms are only expected to evolve if they carry lower costs than harbouring an infection. Mounting an immune response, coping with immunopathological and parasite-induced damage, harbouring high loads of parasites, alteration of species interactions, and reduced survival and fecundity are costs that can differ between responding and non-responding individuals. Encapsulation of *A. crassus* has been observed in both novel and native eel hosts. How encapsulation influences infection intensity, host health, and interactions with other parasites has not been determined yet. Thus, the question for the second part of the book was whether encapsulation is associated with reduced *A. crassus* load and reduced metabolic costs in the European eel. Furthermore, the book intended to determine if encapsulating *A. crassus* changes the parasite community of European eels.

Chapters I and II focus on response processes in experimentally infected novel and native eel hosts and chapter III focuses on consequences of encapsulation for the European eel in the natural environment. For **chapter I**, the aim was to identify the response of the European eel to the migrating phase of invading *A. crassus* larvae, because this phase is proposed to be critical for establishment of helminth parasites. The primary goal was to determine if the European eel elicits a systemic immune response. Additionally, non-immune processes that contribute to the response to *A. crassus* were to be identified. To achieve this, transcriptome-wide gene expression estimates of experimentally infected and uninfected control eels were used to describe expression changes in two major immune organs at 3 days post-infection (dpi). For **chapter II**, the aim was to identify differences in the gene expression response of *A. crassus*' native and novel host. Specifically, the aim was to determine if immune and non-immune responses elicited by larval *A. crassus* follow the pattern observed in other vertebrate-invasive parasite systems, and if they can account for the different susceptibilities between the two host species. This would provide support for the hypothesis that effective defence strategies are missing in novel hosts. Transcriptome-wide gene expression in the head kidney, an immune organ in fish, was used to identify genes and processes that were modified upon infection. Both immune and non-immune processes during the early stages of infection (3 and 23 dpi) and their contribution to the different infection outcomes between host species were measured. For **chapter III**, the aim was to determine if encapsulation of *A. crassus*, a presumably evolved response, can successfully constrain infections in the European eel. A second goal was to determine if encapsulation is associated with a specific immune response and if it reduces metabolic costs in the continental yellow eel stage. Lastly, the effect of encapsulation on the parasite community composition were to be determined. For this, *A. crassus* infection intensity and parasite community composition were compared between responding, i.e. encapsulating, and non-responding, i.e. non-encapsulating, European eels. Furthermore, expression differences of immune and metabolic candidate genes, identified from the transcriptome-wide expression studies (chapters I and II), were determined between responding and non-responding individuals.

Chapter I

Experimental infection with *Anguillicola crassus* alters immune gene expression in both spleen and head kidney of the European eel (*Anguilla anguilla*)

Seraina E. Bracamonte[1,2,3,*], Paul R. Johnston[1,2,4], Klaus Knopf[1,3], Michael T. Monaghan[1,2,4]

[1]Leibniz-Institute of Freshwater Ecology and Inland Fisheries, Müggelseedamm 301/310, 12587 Berlin, Germany
[2]Berlin Center for Genomics in Biodiversity Research, Königin-Luise-Strasse 6-8, 14195 Berlin, Germany
[3]Humboldt-Universität zu Berlin, Invalidenstrasse 42, 10115 Berlin, Germany
[4]Freie Universität Berlin, Königin-Luise-Strasse 1-3, 14195 Berlin, Germany

Published in Marine Genomics 45 (2019) pp. 28-37. doi: 10.1016/j.margen.2018.12.002

Introduction

Recent decades have seen an increase of newly emerging diseases all over the world. One reason for this is the accidental introduction of non-native parasites by humans (Daszak et al., 2000, Peeler et al., 2011). These parasites can be highly virulent in new hosts and there are many examples of translocated parasites posing a considerable threat to their new host species (Dunn & Hatcher, 2015, Smith et al., 2009). In European fresh waters, invasive parasites have been implicated in the decimation of fish, amphibian, and invertebrate populations (Peeler et al., 2011, Gozlan et al., 2005, McKenzie & Peterson, 2012, Kozubikova et al., 2007). In many cases the reason for the high susceptibility of naïve hosts and the physiological processes underlying the response to invasive parasites are poorly understood.

The introduction of the parasite *Anguillicola crassus* Kuwahara, Niimi & Hagaki, 1974 into Europe is considered to be one of the factors that has led to the decline of the European eel (*Anguilla anguilla* L.) population (Kirk, 2003). *A. crassus* is a trophically transmitted swim bladder nematode native to the Japanese eel (*A. japonica* Temminck & Schlegel, 1846). It was introduced into Europe in the early 1980s from Taiwan (Wielgoss et al., 2008), first detected in wild eels in Germany in 1982, and reported from other countries soon afterward (Peters & Hartmann, 1986, Kirk, 2003). It has now reached a prevalence of 50-90% across most of the distribution range of the European eel (ICES, 2012, Lefebvre & Crivelli, 2004).

Several effects of *A. crassus* infection have been documented in infected European eels, including histopathological changes of the swim bladder wall and altered gas composition in the swim bladder which likely affects its functioning (Würtz et al., 1996, Würtz & Taraschewski, 2000). Although some studies have reported no adverse effects of *A. crassus* on performance of European eels during the freshwater stage of their life cycle under normal conditions (Lefebvre et al., 2013, Kelly et al., 2000), additional stressors such as hypoxia have strong effects on the viability of infected eels and can increase mortality (Lefebvre et al., 2007, Molnár et al., 1991, Gollock et al., 2005). *A. crassus* infections also appear to accelerate and interfere with the silvering process of European eels, which prepares them for the long distance spawning migration across the ocean (Fazio et al., 2012, Pelster et al., 2016). This occurs in combination with increased energy consumption and behavioural alterations during extensive swimming (Palstra et al., 2007, Newbold et al., 2015). *A. crassus* infections might therefore be costly, reducing fitness and quality of spawners and hampering successful completion of migration and reproduction.

A parasite that has a strong effect on the fitness of its host should favour adaptation by the host and therefore the development of a specific immune response. There is some evidence for adaptation by the parasite, as infection intensity and size of European *A. crassus* differed from those of Taiwanese *A. crassus* in experimental infections (Weclawski et al., 2013, Weclawski et al., 2014). On the other hand, there is only little evidence that European eels may be adapting to *A. crassus*. Although infection intensities have stabilized, they are still high (Lefebvre & Crivelli, 2004, Audenaert et al., 2003) and there are only occasional reports of European eels confining the parasite (Audenaert et al., 2003, Molnár, 1994). Immune responses differ markedly among individuals of *A. anguilla*, with some individuals exhibiting a strong response and others not responding at all (Molnár, 1994, Knopf et al., 2000). In cases of host responsiveness, inflammatory cells infiltrate the swim bladder following infection. Within the same host, some *A. crassus* larvae were encapsulated and necrotized while others were not (Molnár, 1994), and the infiltrating cells appeared primarily to remove cellular debris caused by the migrating activity of the larvae (Würtz & Taraschewski, 2000). Antibody production varies considerably among individuals and appears to be elicited by adult nematodes rather than the invading larval stage (Knopf et al., 2000). With respect to immune response, little is known regarding alterations induced at the molecular level by the parasite, and few studies have examined changes in gene expression after *A. crassus* infection. Using a qPCR approach, Fazio et al. (2009, 2012) found indications that the expression of genes involved in osmoregulation, haematopoiesis, and silvering were altered in experimentally infected European eels. Gene expression of a range of processes was differentially regulated in the swim bladder during silvering of naturally infected versus uninfected eels, affecting the modifications necessary for long distance migration in open waters (Pelster et al., 2016). Naturally infected eels also regulated expression of genes associated with swim bladder functioning and with the immune response (Schneebauer et al., 2017). The expression of several inflammatory genes was altered, thus providing additional evidence that a localized immune response can develop in the swim bladder once *A. crassus* has established.

We aimed to identify changes in gene expression and infer the corresponding processes during *A. crassus* infections. We were particularly interested in identifying possible systemic immune responses against the parasite, therefore we focussed on two immune organs of fish, the spleen and head kidney. The spleen is a major secondary lymphoid organ (Whyte, 2007, Ellis, 1980) and it plays an important role in the progression of innate and adaptive immune responses locally and systemically (Bronte & Pittet, 2013). It is involved in clearing the blood and retains antigens for long periods of time. It is also an important site for the destruction of erythrocytes and it functions as metabolic dump (Press & Evensen, 1999). The head kidney is a fish-specific organ equivalent to the bone marrow of higher vertebrates (Whyte, 2007, Alvarez-Pellitero, 2008). It is both a primary and secondary immune organ. As such it is the main site of haematopoiesis. It is important for initiation and progression of an immune response by trapping, processing, and presenting antigens to lymphocytes which trigger an adaptive immune response and it is the main site of antibody production.

We sequenced total mRNA of spleen and head kidney tissue from European eels that were experimentally infected with *A. crassus* and compared expression levels with those from uninfected control eels at three days post-infection (dpi). We hypothesized that this approach would allow us to determine if invading larvae initiate a systemic immune response which is a necessary step for mounting a protective response. While an early time point may not capture the fully mounted immune response, we were interested in the tissue migrating phase of parasites that is considered crucial for their establishment (Mulcahy et al., 2005). We selected 3 dpi because *A. crassus* larvae were reported in the swim bladder as early as 4 dpi (Knopf et al., 1998) and we wanted to capture the migrating phase that is considered crucial for the establishment of a helminth. The response elicited by that stage, or a lack thereof, may hint at why European eels have a low capacity of confining *A. crassus* infections.

Materials and Methods

Infection and sampling

European eels were purchased in 2004 from a commercial eel farm that is free of *A. crassus* (Domäne Voldagsen, Einbeck, Germany). Experimental infections were carried out in November 2014. Eels were kept individually in 40-l compartments within 200-l tanks in a recirculating system with aerated tap water at a water temperature of 22 °C prior to, and during, the experiment. At the beginning of the experiment, five treatment eels were infected with *A. crassus* and five control eels were sham-infected following Knopf et al. (1998). In short, second stage larvae (L$_2$) of *A. crassus* were isolated from swim bladders of infected wild eels (*A. anguilla*) caught in Lake Müggelsee, Germany, in October 2014 and fed to wild-caught copepods from the same lake. Nineteen to 23 days post-infection (dpi), third stage larvae (L$_3$) were isolated from the copepods using the potter method (Haenen et al., 1994). Each of the treatment eels was administered 25 randomly selected L$_3$ suspended in 100 µl of PBS, pH 7.2 with a stomach tube. The same amount of PBS containing no *A. crassus* larvae was administered to control eels. Eels were decapitated 3 dpi, when the parasite is migrating from the intestine to the swim bladder (Knopf et al., 1998). The spleen and the head kidney (defined according to Tesch, 2003) were removed and immediately stored in RNA*later*® (Life Technologies, Darmstadt, Germany) following the manufacturer's recommendations. Tissues were then kept in RNA*later*® at -20 °C until extraction of RNA. The swim bladders of all eels were examined and all *A. crassus* were counted. Encapsulated *A. crassus* were not present. At the end of the experiment, eels weighed 115.9 ± 24.8 g. The study was approved by the Berlin State Office for Health and Social Affairs (LaGeSo) in Berlin, Germany (approval number G 0021/15).

RNA extraction and sequencing

RNA was extracted from the spleen and the head kidney using TRIzol® Reagent (Life Technologies, Darmstadt, Germany) following the manufacturer's recommendations for fatty tissue with slight modifications as described below. Each tissue type was homogenized in 850 µl of TRIzol twice for 1.5 min at 18/s in a TissueLyser II (Qiagen, Hilden, Germany) and centrifuged at 12,000 x g for 10 min at 4 °C. An additional 150 µl of TRIzol was added to the aqueous phase prior to chloroform (200 µl) addition. RNA was precipitated from the aqueous phase with 500 µl isopropanol. The pellet was washed with 1 ml 70% ethanol, dried on a heat block for 5 min at 28 °C, and resuspended in 50 µl DEPC water (Life Technologies). RNA was then incubated on a heat block for 2 min at 50 °C. RNA quality and quantity were determined using a 2100 Bioanalyzer (Agilent Technologies, Santa Clara, USA). The samples were diluted to a concentration of 40 ng/µl in TE buffer pH 7.5 and shipped on dry ice to Beckman Coulter Genomics (Danvers, USA) for mRNA library preparation, paired-end (100 bp) sequencing on an Illumina HiSeq 2000, and quality control. The spleen and the head kidney of six samples (3 infected, 3 control) were sequenced in one run (batch A, $n = 3$), while another three spleen samples (2 infected, 1 control) and four head kidney samples (2 infected, 2 control) were sequenced in a different run (batch B, $n = 2$). RNA for the two batches of samples was extracted at different time points but using the same procedure.

Transcriptome assembly and annotation

A single *de novo* assembly was produced using reads from all samples of both tissues with Phred scores > 30. Reads were assembled using Trinity v2.1.1 (Grabherr et al., 2011, Haas et al., 2013) with the default settings. The Trimmomatic option for trimming low quality reads was used with the default settings (SLIDINGWINDOW:4:5 LEADING:5 TRAILING:5 MINLEN:25). In silico normalization was applied to restrict the maximum read coverage to 50. The transcriptome was annotated by blastx searches against the UniProtKB/Swiss-Prot (www.uniprot.org) and RefSeq (www.ncbi.nlm.nih.gov/refseq/) databases. The E-value cut-off for both was set to 1e-3. Trinotate v3.0.0 (https://trinotate.github.io/) was used to obtain corresponding Gene Ontology (GO) assignments for UniProtKB/Swiss-Prot-derived annotations. The taxonomic composition for best matches obtained from RefSeq was examined to check for signs of obvious contamination with non-eel RNA.

Differential gene expression and functional analysis

Different tissues are known to respond differently to parasite infections (Huang et al., 2016, Skugor et al., 2008, Robledo et al., 2014), therefore we contrasted gene expression in the spleen and the head kidney between infected and control eels separately for each tissue. Reads were re-aligned separately for each of the 9 spleen and 10 head kidney samples using the assembled transcriptome as a reference for abundance estimates using RSEM v1.2.26 and the script provided by Trinity. These estimates were used for calculating differential gene expression between infected and control eels separately for the spleen and the head kidney with DESeq2 v1.10.1 (Love et al., 2014). Aside from the treatment, the sequencing run was included into the model to control for batch (A, B, see above) effects. Genes with a mean coverage below 10 (low coverage) were excluded from the analysis, as recommended by Todd et al. (2016), and a local fit was used to estimate the dispersions. Genes were considered significantly differentially expressed if they had an adjusted p-value < 0.05 and a $\log_2$ fold change $\geq \pm 1$. Adjustment of p-values followed the Benjamini-Hochberg procedure for multiple testing (Benjamini & Hochberg, 1995) implemented in DESeq2. A principal component analysis (PCA) was done to determine the overall effect of treatment on sample relationship within tissues and the differences between the spleen and the head kidney. For the PCA, samples of both tissues were combined into one dataset which only included genes that passed the mean coverage filter (see above) in both tissues. The prcomp function in R (R Core Team, 2016) was used on the variance-stabilized counts of the RSEM abundance estimates to perform the PCA.

Enriched GO terms were identified by conditional hypergeometric tests using the GOstats package v2.36.0 (Falcon & Gentleman, 2007) for R. Custom GO annotations obtained from the Trinotate annotation of the transcriptome were used as reference for the GO terms of the differentially expressed genes (DEG) to be compared. The analysis was restricted to GO terms from the domain "biological processes". The significance level for enriched GO terms was set to 0.01.

Results

All eels that were administered *A. crassus* had living L_3 in their swim bladders at the end of the experiment. Mean infection intensity was 2.4 parasites per eel (range: 1-3). More advanced parasitic stages were not present and the majority of larvae were likely still migrating from the intestine to the swim bladder. None of the control eels were infected.

Transcriptome assembly and annotation

Sequencing of spleen and head kidney mRNA resulted in > 860 M reads (438,649,866 reads in batch A and 423,602,528 reads in batch B; see Materials and Methods section). Of these, 95.3% and 96.0% passed quality control (QC). The mean number of high quality reads per sample was 34.8 M in batch A and 50.8 M in batch B. There were no significant differences in number or quality of reads between spleen and head kidney samples. All reads that passed QC were assembled into 578,084 contigs and 823,359 isoforms with an average length of 620 bases and an N50 of 948 bases. The GC content for these sequences was 46.03%. UniProt annotations were assigned to 74,796 contigs (13%

of the total genes) and RefSeq annotations were assigned to 50,311 contigs (8.7%). Twelve of the top 15 matches from RefSeq were from fish genera. Most matches came from *Lepisosteus*, *Danio*, and *Astyanax* and these together provided best matches for about 44% of all annotated contigs.

Differential gene expression

Excluding contigs with a low coverage (see Materials and Methods section) reduced the number that was used for the analysis to 59,666 in the spleen and 63,672 in the head kidney. Of these, 49.5% and 50.9%, respectively, had UniProt annotations. We refer to those contigs with sufficient coverage as genes throughout the rest of the manuscript. Treatment and control samples clustered together in a tissue-specific manner in the PCA (Fig. I.1), indicating that the spleen and the head kidney differed considerably in their expression profiles regardless of the infection status. Within tissues, samples did not separate into control and treatment groups when considering the expression profiles of all genes. One head kidney sample of the control group and one head kidney sample of the treatment clustered separately from all other head kidney samples (Fig. I.1). GO enrichment of DEG between those two samples and the rest did not allow us to identify the reason for the differences. Therefore, they were included in all further analyses.

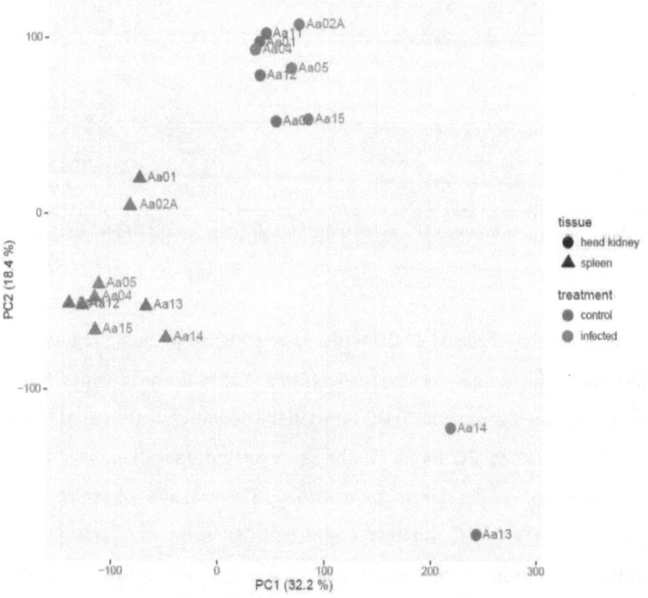

Figure I.1 Principal component analysis (PCA) separating head kidney (●) from spleen (▲) samples but not infected (turquoise) from control (purple) samples within tissues. Head kidney and spleen samples from the same individual are labelled identically.

The number of DEG between infected and control individuals was different in the two tissues. In the spleen, 67 genes were differentially expressed and in the head kidney, 257 genes were differentially expressed (Table SI.1). Of these, 32 in the spleen and 134 in the head kidney had either UniProt or RefSeq annotations. In the spleen, the number of up- and down-regulated genes was very similar (32 vs 35; Fig. I.2a). In the head kidney more genes were up-regulated (196) than down-regulated (61; Fig. I.2b). The $\log_2$ fold changes ranged from -4.9 to 2.9 in the spleen and from -5.3 to 4.8 in the head kidney. Only 13 genes were differentially expressed in both tissues. Of these, 7 were up-regulated and 6 were down-regulated in both spleen and head kidney (Table SI.1). Four of each were annotated.

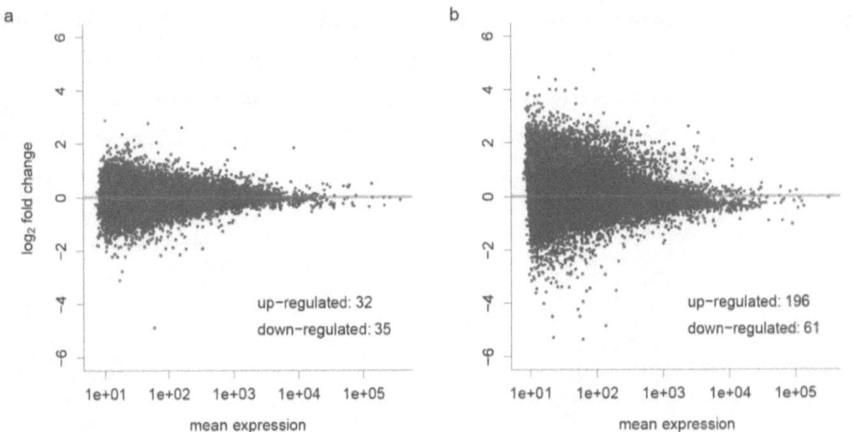

Figure I.2 MA-plots showing differential gene expression between infected and controls for (a) the spleen and (b) the head kidney. Differentially expressed genes are shown as red dots and the number of up- and down-regulated genes is indicated.

Functional analysis

The enrichment analysis revealed 32 GO terms to be enriched in the up-regulated genes and 20 GO terms in the down-regulated genes in the spleen (Table SI.2). In the head kidney, 99 GO terms were enriched in the up-regulated genes and 20 GO terms in the down-regulated genes (Table SI.3). In the spleen, 18 out of 32 enriched GO terms for the up-regulated genes belonged to the categories "immune system process" or "response to stimulus", the majority of which were related to inflammatory processes (Table I.1). Another 4 enriched GO terms were related to regulation of cytokine biosynthesis and production. Furthermore, "leukotriene biosynthetic process" was enriched (Table SI.2). Among the enriched GO terms for down-regulated genes 3 belonged to the categories "immune system process" or "response to stimulus" and 3 enriched GO terms were related to the

Table I.1 Enriched GO terms in the spleen related to "immune system process" or "response to stimulus".

Category	Go term (biological process)	Expected	Count	Size	Direction	P-value
GO:0006953	acute-phase response	0	2	119	↑	<0.001
GO:0001788	antibody-dependent cellular cytotoxicity	0	1	3	↑	<0.001
GO:0001805	positive regulation of type III hypersensitivity	0	1	6	↑	0.001
GO:0019884	antigen processing and presentation of exogenous antigen	0	2	360	↑	0.001
GO:0006954	inflammatory response	0	3	1754	↑	0.001
GO:0071493	cellular response to UV-B	0	1	13	↑	0.002
GO:0071492	cellular response to UV-A	0	1	14	↑	0.002
GO:0001798	positive regulation of type IIa hypersensitivity	0	1	14	↑	0.002
GO:0002892	regulation of type II hypersensitivity	0	1	14	↑	0.002
GO:0048002	antigen processing and presentation of peptide antigen	0	2	543	↑	0.002
GO:0042742	defense response to bacterium	0	2	550	↑	0.002
GO:0001812	positive regulation of type I hypersensitivity	0	1	21	↑	0.003
GO:0002866	positive regulation of acute inflammatory response to antigenic stimulus	0	1	33	↑	0.004
GO:0002524	hypersensitivity	0	1	39	↑	0.005
GO:0043306	positive regulation of mast cell degranulation	0	1	44	↑	0.006
GO:0002696	positive regulation of leukocyte activation	0	2	962	↑	0.006
GO:0002861	regulation of inflammatory response to antigenic stimulus	0	1	49	↑	0.006
GO:0042590	antigen processing and presentation of exogenous peptide antigen via MHC class I	0	1	76	↑	0.010
GO:0002474	antigen processing and presentation of peptide antigen via MHC class I	0	2	342	↓	0.002
GO:1990523	bone regeneration	0	1	22	↓	0.004
GO:0019882	antigen processing and presentation	0	2	651	↓	0.007

Expected and Count give the number of expected and observed differentially expressed genes assigned to the respective category. Size is the total number of genes in the reference transcriptome assigned to that category. Direction indicates enrichment in up-regulated (↑) or down-regulated (↓) genes in infected individuals.

regulation of chemokine production and secretion. Furthermore, processes involved glycopeptide and carbohydrate metabolism were enriched (Tables I.1 & SI.2).

In the head kidney, none of the 99 enriched GO terms for the up-regulated genes belonged to the category "immune system process" and only 11 were subcategories of "response to stimulus", 4 of them relating to DNA damage response and cellular senescence (Table I.2). Additionally, GO terms related to the regulation of arachidonic acid and icosanoid secretion were enriched. Also, cell adhesion processes and anatomical structure development, mainly related to kidney development, were enriched (Table SI.3). For the down-regulated genes 15 enriched GO terms were subcategories of "immune system process". Of the remaining 5 GO terms, 4 belonged to the category "response to stimulus" (Table I.2). Enriched GO terms indicate that immunoglobulin- and lymphocyte-mediated immune responses as well as Fc receptor signalling were down-regulated.

Closer examination of the functions of DEG revealed that 11 in the spleen and 26 in the head kidney were classified as genes of immune system process or response to stimulus according to UniProt annotations (Tables I.3 & I.4). Among them, up- and down-regulated genes in the spleen were involved in inflammation, cell migration, and differentiation and activation of macrophages, mast cells, and lymphocytes. In the head kidney, the up-regulated genes were mostly associated with the cytoskeleton

Table I.2 Enriched GO terms in the head kidney related to "immune system process" or "response to stimulus".

Category	Go term (biological process)	Expected	Count	Size	Direction	P-value
GO:0007185	transmembrane receptor protein tyrosine phosphatase signaling pathway	0	3	46	↑	<0.001
GO:0046426	negative regulation of JAK-STAT cascade	0	4	163	↑	<0.001
GO:0090403	oxidative stress-induced premature senescence	0	2	16	↑	<0.001
GO:0043517	positive regulation of DNA damage response, signal transduction by p53 class mediator	0	2	26	↑	0.001
GO:0009968	negative regulation of signal transduction	5	13	3839	↑	0.001
GO:0090398	cellular senescence	0	3	158	↑	0.001
GO:0030520	intracellular estrogen receptor signaling pathway	0	3	180	↑	0.002
GO:0042770	signal transduction in response to DNA damage	0	3	275	↑	0.005
GO:0071307	cellular response to vitamin K	0	1	6	↑	0.008
GO:0051387	negative regulation of neurotrophin TRK receptor signaling pathway	0	1	7	↑	0.009
GO:0052697	xenobiotic glucuronidation	0	1	7	↑	0.009
GO:0045087	innate immune response	1	6	2927	↓	<0.001
GO:0002474	antigen processing and presentation of peptide antigen via MHC class I	0	3	342	↓	<0.001
GO:0019882	antigen processing and presentation	0	3	651	↓	<0.001
GO:0006958	complement activation, classical pathway	0	2	342	↓	0.002
GO:0002757	immune response-activating signal transduction	0	3	1532	↓	0.002
GO:0038096	Fc-gamma receptor signaling pathway involved in phagocytosis	0	2	394	↓	0.002
GO:0072376	protein activation cascade	0	2	430	↓	0.002
GO:0002440	production of molecular mediator of immune response	0	2	458	↓	0.003
GO:0016064	immunoglobulin mediated immune response	0	2	513	↓	0.004
GO:0042742	defense response to bacterium	0	2	550	↓	0.004
GO:0034165	positive regulation of toll-like receptor 9 signaling pathway	0	1	25	↓	0.004
GO:0050778	positive regulation of immune response	0	3	2038	↓	0.005
GO:0006959	humoral immune response	0	2	620	↓	0.005
GO:0002682	regulation of immune system process	1	4	4405	↓	0.005
GO:0002449	lymphocyte mediated immunity	0	2	795	↓	0.008
GO:0051707	response to other organism	0	3	2536	↓	0.009
GO:0002460	adaptive immune response based on somatic recombination of immune receptors built from immunoglobulin superfamily domains	0	2	830	↓	0.009
GO:0038095	Fc-epsilon receptor signaling pathway	0	2	844	↓	0.009
GO:0009607	response to biotic stimulus	0	3	2602	↓	0.009

Columns as described for Table I.1.

and extracellular matrix. Also genes involved in B cell maturation and inflammation were up-regulated. Down-regulated genes in the head kidney were involved in immunoglobulin formation and T cell activation.

The vast majority of annotated DEG in the spleen and down-regulated genes in the head kidney was related to immune response or the regulation thereof, although they are not classified as genes of immune system process or response to stimulus by UniProt annotations (Tables I.3, I.4 & SI.1). They were associated with inflammation, cell proliferation, activation, and migration, and wound healing. In the head kidney several up-regulated genes that were not classified as genes of immune system process or response to stimulus were implicated in wound healing, B cell development, and cell migration as well (Table SI.1).

Ferrochelatase, an enzyme of the heme biosynthesis pathway, and the scavenger receptor cysteine-rich type 1 protein M130, which is involved in haemoglobin scavenging, were both up-regulated in the spleen. In the head kidney, several genes involved in osmoregulation and paracellular flow were upregulated, among them aquaporin-5, claudin-8, and multiple PDZ domain protein 1.

Table I.3 Differentially expressed genes in the spleen associated with the immune system.

Gene ID	Log2FC	Wald stat	P-value	Adj. p-value	Direction	Uniprot and refseq annotations
TRINITY_DN168651_c5_g8	2.14	5.97	2.35E-09	2.26E-05	↑	High affinity immunoglobulin gamma Fc receptor I
TRINITY_DN164847_c1_g5	2.37	5.56	2.71E-08	1.96E-04	↑	HLA class II histocompatibility antigen, DR beta 4 chain
TRINITY_DN85147_c0_g2	1.99	4.67	3.06E-06	0.006	↑	carboxypeptidase N subunit 2-like
TRINITY_DN172142_c2_g9	1.62	4.68	2.84E-06	0.006	↑	Scavenger receptor cysteine-rich type 1 protein M130 (CD163)
TRINITY_DN172142_c2_g4	1.53	4.36	1.31E-05	0.017	↑	Scavenger receptor cysteine-rich type 1 protein M130 (CD163)
TRINITY_DN160149_c1_g6	1.66	4.27	1.94E-05	0.023	↑	Ig heavy chain V-I region HG3
TRINITY_DN150174_c4_g2	1.78	4.17	3.01E-05	0.031	↑	Neprilysin
TRINITY_DN131224_c0_g2	1.33	4.05	5.06E-05	0.044	↑	Cathelicidin
TRINITY_DN137702_c1_g8	-4.88	-12.80	1.56E-37	9.00E-33	↓	Major histocompatibility complex class I-related gene protein
TRINITY_DN162543_c1_g4	-3.11	-7.48	7.62E-14	2.20E-09	↓	CD48 antigen
TRINITY_DN172627_c9_g2	-1.89	-5.38	7.44E-08	4.31E-04	↓	Interferon-induced very large GTPase 1
TRINITY_DN158064_c3_g2	-1.70	-5.23	1.72E-07	7.68E-04	↓	Early growth response protein 1
TRINITY_DN170110_c2_g11	-2.10	-4.93	8.02E-07	0.002	↓	Periostin
TRINITY_DN165611_c2_g3	-1.86	-4.42	9.75E-06	0.014	↓	Major histocompatibility complex class I-related gene protein
TRINITY_DN139249_c2_g10	-1.82	-4.26	2.01E-05	0.024	↓	GTPase IMAP family member
TRINITY_DN162962_c0_g2	-1.69	-4.13	3.61E-05	0.035	↓	Actin-related protein 2/3 complex subunit 1A
TRINITY_DN169065_c0_g2	-1.46	-4.09	4.32E-05	0.038	↓	Early growth response protein 1

Log2FC is the log2 fold change between expression in infected and control individuals, Wald stat is the Wald statistic calculated by DESeq2, Adj. p-value is the Benjamini-Hochberg adjusted p-value, and Direction indicates up-regulation (↑) or down-regulation (↓) in infected individuals. Gene IDs in bold indicate differential expression also in the head kidney.

Discussion

Anguillicola crassus is a parasitic nematode that was introduced into European fresh waters in the 1980s and has acquired the European eel (*Anguilla anguilla*) as a new host (Kirk, 2003). Previous studies of the *A. anguilla* immune response to *A. crassus* have found localized inflammation in the swim bladder (Würtz & Taraschewski, 2000, Schneebauer et al., 2017) and an antibody response against adults (Knopf et al., 2000). We wanted to know whether the European eel is capable of mounting an immune response during the early stages of infections and used transcriptome-wide gene expression in two immune organs of European eels, the spleen and the head kidney, to detect immunological changes occurring 3 days post-infection (dpi) during the larval migrating phase of *A. crassus*. This early stage has been proposed to be critical for establishment of helminthic infections

(Mulcahy et al., 2005). We found expression of genes involved in an immune response to be modified in both organs. Infection also led to modified expression of genes related to heme metabolism, the O_2-binding unit of haemoproteins such as haemoglobin and cytochrome, in the spleen, and of genes involved in osmoregulation and renal function in the head kidney. Therefore, the European eel activates the immune system, but the overall response appears to be more complex. Below we first discuss genes and processes associated with an immune response to the parasite. We then discuss non-immune-related genes and processes that may influence the performance of infected European eels.

Table I.4 Differentially expressed genes in the head kidney associated with the immune system.

Gene ID	Log2FC	Wald stat	P-value	Adj. p-value	Direction	Uniprot and refseq annotations
TRINITY_DN146291_c0_g1	3.14	5.10	3.41E-07	7.09E-04	↑	Estrogen receptor beta
TRINITY_DN164847_c1_g5	3.83	5.08	3.70E-07	7.43E-04	↑	HLA class II histocompatibility antigen, DR beta 4 chain
TRINITY_DN168593_c4_g3	3.01	4.83	1.39E-06	0.002	↑	Estrogen receptor beta
TRINITY_DN136158_c0_g1	1.77	4.55	5.25E-06	0.006	↑	Receptor tyrosine-protein kinase erbB-3
TRINITY_DN158138_c1_g1	2.05	4.55	5.28E-06	0.006	↑	Receptor-type tyrosine-protein phosphatase delta
TRINITY_DN139426_c1_g2	3.03	4.51	6.54E-06	0.007	↑	Biglycan
TRINITY_DN151863_c1_g1	2.34	4.43	9.29E-06	0.008	↑	Collagen alpha-5(IV) chain
TRINITY_DN149909_c1_g3	1.88	4.45	8.75E-06	0.008	↑	Receptor-type tyrosine-protein phosphatase delta
TRINITY_DN163104_c3_g2	2.43	4.37	1.23E-05	0.010	↑	Atrial natriuretic peptide receptor 1
TRINITY_DN157378_c3_g1	2.13	4.38	1.21E-05	0.010	↑	Receptor-type tyrosine-protein phosphatase delta
TRINITY_DN144825_c0_g1	1.63	4.35	1.37E-05	0.011	↑	SH2 domain-containing protein 3A
TRINITY_DN159222_c0_g1	1.76	4.32	1.57E-05	0.011	↑	Platelet-derived growth factor C
TRINITY_DN171723_c2_g4	1.97	4.24	2.25E-05	0.015	↑	Receptor-type tyrosine-protein phosphatase delta
TRINITY_DN162605_c1_g2	1.68	4.21	2.55E-05	0.016	↑	Wilms tumor protein
TRINITY_DN148691_c0_g1	2.52	4.20	2.68E-05	0.016	↑	Receptor-interacting serine/threonine-protein kinase 4
TRINITY_DN169331_c3_g2	1.40	4.17	3.00E-05	0.018	↑	1-phosphatidylinositol 4,5-bisphosphate phosphodiesterase delta
TRINITY_DN164726_c4_g1	1.50	4.15	3.26E-05	0.018	↑	Secretory phospholipase A2 receptor
TRINITY_DN157378_c3_g2	2.04	4.09	4.35E-05	0.021	↑	Receptor-type tyrosine-protein phosphatase delta
TRINITY_DN160149_c1_g6	1.65	4.05	5.02E-05	0.023	↑	Ig heavy chain V-I region HG3
TRINITY_DN128669_c0_g2	1.90	4.05	5.19E-05	0.023	↑	Serine/threonine-protein kinase pim-3
TRINITY_DN156836_c3_g2	1.42	4.05	5.15E-05	0.023	↑	Receptor-type tyrosine-protein phosphatase delta
TRINITY_DN171512_c0_g1	2.44	4.04	5.45E-05	0.024	↑	Desmoplakin
TRINITY_DN152103_c1_g1	1.65	4.01	6.05E-05	0.026	↑	finTRIM family, member 82
TRINITY_DN153657_c2_g1	2.15	3.97	7.13E-05	0.028	↑	Collagen alpha-5(IV) chain-like
TRINITY_DN161130_c4_g1	2.54	3.95	7.77E-05	0.029	↑	Estrogen receptor beta
TRINITY_DN133663_c0_g1	2.67	3.94	7.99E-05	0.029	↑	Microtubule-actin cross-linking factor 1, isoforms 1/2/3/5
TRINITY_DN127233_c1_g1	2.91	3.94	7.98E-05	0.029	↑	Sex hormone-binding globulin
TRINITY_DN127456_c0_g1	2.65	3.90	9.46E-05	0.032	↑	Microtubule-actin cross-linking factor 1, isoforms 1/2/3/5
TRINITY_DN119036_c0_g1	2.83	3.87	1.08E-04	0.035	↑	Ras association domain-containing protein 5

Table I.4 continued.

Gene ID						Description
TRINITY_DN119036_c0_g1	2.83	3.87	1.08E-04	0.035	↑	Ras association domain-containing protein 5
TRINITY_DN142758_c0_g1	2.36	3.85	1.16E-04	0.036	↑	basement membrane-specific heparan sulfate proteoglycan core protein-like
TRINITY_DN156129_c0_g2	2.17	3.83	1.26E-04	0.037	↑	UDP-glucuronosyltransferase 1
TRINITY_DN149035_c4_g1	1.71	3.77	1.65E-04	0.041	↑	Beta-1,4 N-acetylgalactosaminyltransferase 2
TRINITY_DN165320_c4_g1	2.36	3.74	1.81E-04	0.044	↑	Protein kinase C delta type
TRINITY_DN145200_c1_g2	2.71	3.73	1.95E-04	0.046	↑	Desmoglein-1
TRINITY_DN142233_c0_g1	1.58	3.72	2.01E-04	0.046	↑	Secretory phospholipase A2 receptor
TRINITY_DN166241_c2_g1	1.89	3.69	2.21E-04	0.048	↑	Homeodomain-interacting protein kinase 2
TRINITY_DN137702_c1_g10	-5.33	-7.64	2.14E-14	6.24E-10	↓	Major histocompatibility complex class I-related gene protein
TRINITY_DN137702_c1_g8	-5.28	-7.68	1.63E-14	6.24E-10	↓	Major histocompatibility complex class I-related gene protein
TRINITY_DN155813_c1_g1	-2.58	-5.89	3.80E-09	1.84E-05	↓	BTB/POZ domain-containing protein 17
TRINITY_DN168857_c2_g5	-2.86	-4.91	9.28E-07	0.002	↓	GTPase IMAP family member 4
TRINITY_DN162543_c1_g4	-3.61	-4.81	1.51E-06	0.002	↓	CD48 antigen
TRINITY_DN152464_c2_g16	-3.15	-4.34	1.43E-05	0.011	↓	Granulins
TRINITY_DN120453_c1_g2	-1.90	-4.16	3.14E-05	0.018	↓	Ig heavy chain V-III region 23
TRINITY_DN144819_c0_g4	-1.60	-4.13	3.64E-05	0.019	↓	P-selectin
TRINITY_DN144089_c4_g6	-1.53	-4.09	4.39E-05	0.021	↓	Ig lambda chain V-V region DEL
TRINITY_DN128201_c0_g3	-3.05	-4.00	6.45E-05	0.026	↓	CMRF35-like molecule
TRINITY_DN152706_c1_g13	-2.58	-3.89	1.02E-04	0.034	↓	GTPase IMAP family member 4
TRINITY_DN156202_c2_g3	-2.91	-3.85	1.18E-04	0.036	↓	IgGFc-binding protein
TRINITY_DN147923_c2_g2	-2.66	-3.81	1.38E-04	0.038	↓	Myosin-11
TRINITY_DN165611_c2_g3	-2.35	-3.80	1.42E-04	0.038	↓	Major histocompatibility complex class I-related gene protein
TRINITY_DN168771_c3_g3	-2.90	-3.78	1.57E-04	0.040	↓	Ig kappa chain V-V region HP 91A3
TRINITY_DN164893_c1_g8	-2.17	-3.78	1.60E-04	0.040	↓	HERV-H LTR-associating protein 2-like
TRINITY_DN151002_c1_g1	-1.55	-3.74	1.85E-04	0.044	↓	High affinity immunoglobulin epsilon receptor subunit beta-like

Columns as described for Table I.3. Gene ID in bold indicates that this gene was also differentially expressed in the spleen.

Immune-related processes

There were several indicators in both organs that *A. crassus* elicited an immune response in *A. anguilla*. Specifically, genes of the major histocompatibility complex class II (MHC II) were up-regulated in both organs. MHC II displays parasitic antigens that are recognized by the T cell receptor (TCR; Morris et al., 1994). Successful binding to MHC II-antigen complexes activates helper T cells and induces a specific immune response (Murphy, 2012). European eels express up to four different MHC IIB variants (Bracamonte et al., 2015). Whether up-regulation of MHC IIB gene expression implies presentation of *A. crassus*-derived antigens and increased TCR signalling cannot be determined with our data, but suggests some immune response. Genes that modulate inflammatory processes were modified in both head kidney and spleen. The phospholipase A2 receptor gene (*pla2r*) was up-regulated in the head kidney. Ligand binding to PLA2R induces the production of pro-inflammatory mediators (Granata et al., 2005, Park et al., 2003) and controlled cell death (apoptosis) upon DNA damage through the production of arachidonic acid (Augert et al., 2009, Pan et al., 2014). Cathelicidins, up-regulated in the spleen in this study, are anti-microbial peptides with immunomodulatory properties (Brown & Hancock, 2006). They may promote either a pro-inflammatory or an anti-inflammatory response. Elevated expression of the scavenger receptor cysteine-rich type 1 protein M130 gene (*cd163*) also suggests suppression of inflammation. CD163 is characteristic for alternatively activated macrophages (Van Gorp et al., 2010, Kowal et al., 2011). These macrophages are generally associated with a T helper type 2 (Th2) immune response, suppression of an inflammatory response, and wound healing (Gause et al., 2013). Whether CD163 plays a role in the protective function against helminths has not yet been resolved.

There were also several indicators that *A. anguilla* did not produce an immune response at 3 dpi. The early growth response protein 1 gene (*egr-1*) was down-regulated in the spleen. Egr-1 is a transcription factor that is involved in the activation of lymphocytes, primarily Th2 cells, and the stimulation of interleukin-4 expression (Lohoff et al., 2010), an important regulator of the humoral immune response. It further promotes the differentiation of macrophages and activation of mast cells (McMahon & Monroe, 1996, Li et al., 2006). All of these processes play an important role in an effective immune response in mice and humans against helminths (Gause et al., 2013). Down-regulation of *egr-1* also indicates decreased proliferation of mature B cells (Gururajan et al., 2008). Impairment of B cells, the antibody (Ab)-producing cells, was further supported by the up-regulation of the protein kinase C δ gene (*pkcd*) that we observed in the head kidney of infected individuals. PKCD negatively affects B cell development and proliferation (Limnander et al., 2011, Mecklenbräuker et al., 2002, Miyamoto et al., 2002). Accordingly, genes encoding immunoglobulin chains, which compose the B cell receptor and Ab, were down-regulated in the head kidney. We conclude that Ab production might not be initiated at 3 dpi, in agreement with previous findings showing that European eels produce no Ab against *A. crassus* larvae (Knopf et al., 2000).

An anti-inflammatory Th2 response is thought to have evolved in response to macroparasitic infections (e.g. heminths) to protect the host from excessive damage by the immune response itself. A pro-inflammatory Th1 response is usually produced to fight microparasitic infections (e.g. bacteria; Graham et al., 2005). However, there is evidence that migrating parasitic larvae may be better controlled by a Th1 response prior to chronic infections and that this may prevent their establishment (Mulcahy et al., 2005, Moreau & Chauvin, 2010). Our data did not clearly indicate a Th1 response or a Th2 response. An unresolved mix of Th1 response and Th2 response may be the initial state upon encountering a novel parasite. Selection can then drive the response into a Th1 response, which may prevent establishment of infection, or a Th2 response which may reduce damage while permitting the parasite to establish. Infection intensities of *A. crassus* have stabilized since its introduction in the 1980s (Lefebvre & Crivelli, 2004, Audenaert et al., 2003) and there are signs of adaptation (Weclawski et al., 2013). Monitoring the early stages of the immune response in recent eel generations at a higher resolution may indicate if the T helper response is targeted by selection and which trajectory the European eel will follow to cope with this invasive parasite.

Non-immune-related processes

In addition to immune responses, *A. crassus* infections affected genes related to metabolism of *A. anguilla*. CD163, although associated with inflammatory processes, is best known for its function as scavenger of haemoglobin-haptoglobin complexes (Fabriek et al., 2005). Haemoglobin is released from old and defective erythrocytes. Degradation of heme, the O_2-binding unit of haemoglobin, induces an anti-inflammatory response by alternatively activated macrophages (Fabriek et al., 2005). The increased expression of *cd163* that we observed in infected individuals may therefore indicate higher erythrocyte turnover following *A. crassus* infections. This hypothesis is further supported by our observed up-regulation of the gene encoding ferrochelatase (*fech*), the last enzyme in heme biosynthesis (Layer et al., 2010). Infected eels might have a reduced ability to bind and distribute O_2 and up-regulation of heme biosynthesis may be an attempt to meet the O_2 requirement. Höglund et al. (1992) reported slightly decreased values for haemoglobin in naturally infected eels containing adult blood-sucking *A. crassus* and no differences in numbers of erythrocytes. In contrast, Fazio et al. (2009) found increased expression of the gene encoding haemoglobin α in experimentally infected eels, which is in line with our finding of increased expression of *fech*. Heme is also necessary for cellular respiration (Paoli et al., 2002), which provides energy for metabolic processes. Increased *fech* expression may indicate a higher demand of O_2 by infected individuals due to increased energy consumption. The metabolism of carbohydrates and polypeptides was enriched among down-regulated genes in the spleen. We conclude that the energy balance might be disturbed in eels infected with *A. crassus*.

GO enrichment revealed that renal integrity and osmoregulatory function were affected in the head kidney of infected eels. Genes encoding the tight junction proteins claudin and multiple PDZ domain protein 1 were up-regulated. Tight junctions determine epithelial permeability and regulate paracellular transport of solutes, mainly through claudins (Koval, 2006, Yu, 2015). Aquaporins, also up-regulated in the head kidney of infected eels in our study, form water channels and are important for osmoregulation (Cerda & Finn, 2010, Madsen et al., 2015). Fazio et al. (2009) reported an effect on osmoregulation in the intestine, but not until 8 weeks post-infection. Furthermore, they did not observe an effect on the expression of the aquaporin-encoding gene in the gills, although we observed it to be up-regulated in the head kidney.

Organ differences

The head kidney may be more strongly affected by the infection than the spleen. The number of genes and processes affected was more diverse in the head kidney than in the spleen. The differentially expressed genes in the spleen were almost exclusively associated with the immune system while in the head kidney, they were associated with both immune system processes and physiological processes. The head kidney of eels is considerably different in shape and location from those of other fish (Tesch, 2003) and our data suggests that it might contribute extensively to renal processes in addition to its function as an immune organ. Altered reabsorption of solutes and osmoregulation could therefore indicate that the induction of an immune response by *A. crassus* also interferes with renal excretion and osmotic homeostasis.

Two head kidney samples had expression profiles that differed from those of all other head kidney samples. Although we cannot exclude that the samples were contaminated with trunk kidney tissue or higher amounts of blood during removal of the head kidney tissue, GO enrichment gave no such indication, The genes differing between the two outlier samples and all other samples were enriched for > 2000 GO terms that were associated with very diverse processes. Eels cannot be bred efficiently and the individuals used for the experiment were originally caught from the wild for aquaculture. Also, sex-determination of immature eels is difficult and unreliable. Thus, eels were of unknown sex and genetic background, both of which can have an influence on gene expression. Furthermore, the eels had been held in captivity for a length of time similar to the time span of their continental phase (ca. 10 yr) and were likely close to silvering, i.e. preparing for migration in saltwater and initiating sexual maturation. The silvering process involves major physiological modifications (Tesch, 2003) and if eels were at different developmental stages then this may affect gene expression.

Conclusion

Gene expression results indicated that European eels modify immune processes early during *A. crassus* infections. The mix of Th1 and Th2 processes may be one of the reasons why *A. crassus* is not confined immediately after infection; however, the observed response was complex and affected processes other than those involved in immunity but which may interfere with an immune response. Interestingly, this was more pronounced in the head kidney than in the spleen, supporting a potentially important role of the head kidney in immune response. The molecular modifications that we identified here provide a basis for determining processes that may influence European eel performance following infection.

Data Availability

The Transcriptome Shotgun Assembly project has been deposited at DDBJ/ENA/GenBank under the accession GHAH00000000. The version described in this paper is the first version, GHAH01000000.

CHAPTER II

Gene expression response to a nematode parasite in novel and native eel hosts

Seraina E. Bracamonte[1,2,3], Paul R. Johnston[1,2,4], Michael T. Monaghan[1,2,4,*], Klaus Knopf[1,3]

[1]Leibniz-Institute of Freshwater Ecology and Inland Fisheries, Berlin, Germany

[2]Berlin Center for Genomics in Biodiversity Research, Berlin, Germany

[3]Faculty of Life Sciences, Humboldt-Universität zu Berlin, Berlin, Germany

[4]Institut für Biologie, Freie Universität Berlin, Berlin, Germany

In press in Ecology and Evolution

Introduction

The introduction of non-native parasites into foreign habitats has exposed them to novel host species and has promoted several cases of disease emergence (Daszak et al., 2000, Dobson & Foufopoulos, 2001, Peeler et al., 2011). Novel hosts can be highly susceptible, i.e. suffer from high infection intensities (number of parasites per infected host), severe pathologies, and high fitness costs. Novel infections are leading to population declines and local extinctions of species worldwide (Peeler et al., 2011). The fungal parasites causing chytridiomycosis in amphibians and white-nose-syndrome in North American bats have led to population collapses (Frick et al., 2010, Skerratt et al., 2007), and the parasitic mite *Varroa destructor* is a major driver of honey bee declines (Le Conte et al., 2010). The increased susceptibility that has been observed in some novel hosts may be due to a lack of defence mechanisms which the native hosts had acquired during their shared evolutionary history with the parasite (Mastitsky et al., 2010, Peeler et al., 2011).

A number of species of eels are threatened (Jacoby et al., 2015) and non-native parasites in their freshwater habitat have been proposed as a contributing factor in their decline (Drouineau et al., 2018, Miller et al., 2016, Sures & Knopf, 2004). The parasitic swim bladder nematode *Anguillicola crassus* Kuwahara, Niimi & Hagaki, 1974 was introduced into Europe from Southeast Asia where it is native to the Japanese eel (*Anguilla japonica* Temminck & Schlegel, 1846; Fig. II.1). It was first detected in wild European eels (*A. anguilla* L., 1758) in 1982 and has rapidly spread across most of the European eel's distribution range (Kirk, 2003). In the mid-1990s *A. crassus* was also introduced into the American eel (*A. rostrata* Lesueur, 1817) population (Barse & Secor, 1999). The parasite's introduction into Europe coincides with the onset of a steep decline of the European eel population to recruitment levels < 10% of its pre-1980 level (Bornarel et al., 2017, ICES, 2018, Diekmann et al., 2019).

Figure II.1 The Japanese eel (*Anguilla japonica*) is the native host of *Anguillicola crassus*, a parasitic swim bladder nematode invasive in the European eel (*A. anguilla*).

Natural *A. crassus* infections have not been observed to reduce body condition in the European eel (Lefebvre et al., 2013) or to affect its physiological status (Kelly et al., 2000). However, increased stress and mortality have been reported in parasitized European eels that experienced periods of hypoxia (Gollock et al., 2005, Lefebvre et al., 2007, Molnár et al., 1991), indicating a cumulative negative effect from multiple stressors. Natural and experimental *A. crassus* infections impair swim bladder function (Würtz et al., 1996) and laboratory swimming trials indicated that natural infections increase energy consumption and alter swimming behaviour of the European eel and may thus interfere with the spawning migration and reproduction (Palstra et al., 2007, Newbold et al., 2015, Pelster, 2015, Würtz et al., 1996). For the Japanese eel, body condition was not affected by *A. crassus* infections (Han et al., 2008). No data are available on how infection interacts with environmental stress in the Japanese eel, or whether *A. crassus* affects swimming, energy budget, or fitness.

A number of studies have concluded that *A. crassus* infects the European eel more successfully than the Japanese eel. In natural infections of field-caught yellow eels (the continental freshwater feeding stage of the life cycle), infection intensities and parasite prevalence (proportion of infected hosts) were higher in the European eel (Audenaert et al., 2003, Gérard et al., 2013, Knopf, 2006) than in the Japanese eel (Han et al., 2008, Heitlinger et al., 2009, Münderle et al., 2006). Several infection experiments with European sources of *A. crassus* have also reported higher infection intensities in the European eel compared to the Japanese eel 12 or more weeks post-infection (Knopf & Mahnke, 2004, Knopf & Lucius, 2008). Weclawski et al. (2013) found the Japanese eel to be more efficient at killing the parasite than the European eel over the course of infection, i.e. past 50 days post-infection (dpi), although the infection intensity was higher in the Japanese eel at an early stage (25 dpi) when adult parasites start appearing. Earlier stages of infection have not yet been comparatively studied.

The differences in infection intensities that have been observed several weeks after infection have led to the assumption that the Japanese eel produces a more effective immune response (Knopf, 2006, Taraschewski, 2006); however, evidence for this is scarce. Both eel species develop an antibody response to adult *A. crassus* antigens, but there is no indication that the antibody response is associated with protection (Knopf & Lucius, 2008, Nielsen, 1999). In naturally infected European eels, inflammation and immune cells surrounding parasite larvae have been observed in swim bladders containing both larvae and adults (Molnár et al., 1993, van Banning & Haenen, 1990, Würtz & Taraschewski, 2000). In natural infections, encapsulated larvae can be found at similar proportions in the swim bladder walls of both species (Audenaert et al., 2003, Heitlinger et al., 2009). High infection pressure leads to a massive increase in parasite encapsulation in the Japanese eel (Heitlinger et al., 2009).

Studies of the European eel using RNA-seq to examine gene expression have found that genes involved in an immune response were differentially expressed in the swim bladder (the site of infection) of naturally infected eels (i.e. containing parasites of all stages; Schneebauer et al., 2017), as well as in the spleen and the head kidney (immune organs) very soon after experimental infection (3 dpi, i.e. containing only larvae; Bracamonte et al., 2019). Differential regulation of processes associated with both the innate and the adaptive immune system in immune organs and at the site of infection is a common feature in natural and experimental infections in vertebrates (e.g. Alvarez Rojas et al., 2015, Babayan et al., 2018, Huang et al., 2016) and the gradual shift from the regulation of innate to adaptive immune processes can be observed in gene expression studies (Ehret et al., 2017). Additionally, parasite infections cause differential expression of genes not directly related to an immune response, such as those involved in metabolic processes, tissue repair, or organ function and development (Alvarez Rojas et al., 2015, Babayan et al., 2018, Ronza et al., 2016, Zhang et al., 2017) and this has also been observed in the European eel (Bracamonte et al., 2019, Schneebauer et al., 2017). How *A. crassus* affects gene expression in the Japanese eel, and what processes are modified upon infection, have not yet been determined for any parasitic stage.

RNA-seq studies of infection experiments on a range of species indicate that the number of affected processes, the magnitude of change, and the specific genes involved differ considerably among host species-parasite species systems (e.g. Alvarez Rojas et al., 2015, Haase et al., 2016, Kumar et al., 2015, Zhang et al., 2017). Infections with invasive parasites have consistently induced a more pronounced response in susceptible hosts compared to resistant hosts. In both frogs and toads suffering from chytridiomycosis, a larger number of genes were differentially expressed in susceptible species and they were involved in a more diverse set of processes, including several immune-related and metabolic processes (Eskew et al., 2018, Poorten & Rosenblum, 2016). Similar patterns were observed in bats exposed to the fungus causing white-nose-syndrome (Davy et al., 2017, Field et al., 2015) although the resistant species had cleared the infection at the time of sampling. Finally, more genes were differentially expressed in a more susceptible bee species exposed to *Varroa* mites (Zhang et al., 2010), although the diversity of processes was not reported. For the eel-*Anguillicola* host-parasite system, the processes leading to the different outcome of infection between eel species and the seemingly larger impact on the European eel are still unknown but they may also result from this emerging general pattern.

Here we experimentally infected Japanese eels and European eels with *A. crassus* under controlled conditions. We measured the number of parasites in swim bladders and used RNA-seq to estimate gene expression changes in the head kidney at two time points in the early stages of infection: during the migrating phase of the larval parasite (3 dpi) and after the establishment of larvae in the swim bladder (23 dpi). Our main goal was to test whether processes modified during the early stages of infection contribute to the higher susceptibility of the European eel, the novel host, compared to the Japanese eel, the native host. We also tested whether European eels differentially express a larger number of genes and greater diversity of processes, as potentially predicted by recent observations in amphibians and mammals. If novel hosts produce an ineffective immune response, we expected that maintaining homeostasis, such as metabolism and renal function, would be more problematic for the European eel than for the Japanese eel and that changes in gene expression would be the result.

Materials and Methods

Experimental setup and sampling

Japanese eels were imported as glass eels (transition from marine larval stage to freshwater stage) from Japan in 2006 and raised to the yellow eel stage in the laboratory at the Leibniz-Institute of Freshwater Ecology and Inland Fisheries (Berlin, Germany). The eels have never been exposed to *A. crassus*. European eels were purchased as yellow eels in 2004 from an eel farm in Germany (Domäne Voldagsen, Einbeck) that was free of *A. crassus*. Thereafter, both species were kept in recirculation systems in aerated tap water. Each individual was housed in a separate compartment (40-80 L) within larger aquaria (200 L). Each compartment contained a polyethylene tube for hiding.

The parasite *A. crassus* is trophically transmitted and anguillid eels are the only known final hosts (De Charleroy et al., 1990, Nagasawa et al., 1994). Free-living second stage larvae (L_2) hatch from eggs in fresh water and are consumed by crustacean plankton (Copepoda), the intermediate hosts. In the copepod they molt into third-stage larvae (L_3) which is the infective stage for eels. Eels are infected by feeding on intermediate or paratenic hosts. *A. crassus* L_3 migrate from the intestine to the swim bladder wall in approximately one week (Haenen et al., 1989, Knopf et al., 1998). Two to 3 weeks post-infection, they molt into fourth stage larvae (L_4). At 25 dpi, adults can be present in the swim bladder lumen (Weclawski et al., 2013). They reproduce sexually and the eggs are released into the water.

Infection of eels with *A. crassus* was carried out at the beginning of the experiment following the method of Knopf et al. (1998). In short, *A. crassus* eggs were collected from the swim bladder of wild European eels caught from nearby Lake Müggelsee (Berlin, Germany) in autumn 2014. The L_2 were hatched and fed to copepods from the same lake. Three weeks post-infection, the copepods were crushed to extract L_3 that were then suspended in phosphate-buffered saline (PBS, pH 7.2). For both eel species, 10 individuals were infected with 25 L_3 individuals suspended in 100 µl PBS using a stomach tube, while 9 Japanese eels and 10 European eels were sham-infected with 100 µl PBS and served as controls. At 3 dpi, 5 control and 5 infected individuals of each species were dissected and the head kidney was removed and stored in RNA*later* (Life Technologies, Darmstadt, Germany) at -20 °C. The remaining 9 Japanese eels (5 infected, 4 control) and 10 European eels (5/5) were dissected at 23 dpi and the head kidney was stored in RNA*later* at -20 °C or flash-frozen in liquid nitrogen and stored at -80 °C until processing. For all individuals, the swim bladder was removed during dissection and checked for the presence of *A. crassus* under a binocular. *Anguillicola crassus* individuals were counted and their developmental stage determined (L_3, L_4). No adult *A. crassus* were present at any sampling time.

At the time of dissection, all individuals were weighed and measured. The sex was not determined, because gonads of eels are of an undifferentiated state and cannot be determined microscopically until eels reach an advanced stage of maturity during the spawning migration (Tesch, 2003). The Berlin State Office for Health and Social Affairs (LaGeSo) in Germany approved the experimental procedure (approval number G 0021/15).

RNA extraction and sequencing

RNA of European eel samples collected at 3 dpi was extracted and sequenced as described by Bracamonte et al. (2019). Japanese eel samples collected at 3 dpi and 2 control and 2 treatment samples of the European eel collected at 23 dpi were stored in RNA*later*. The remaining 6 samples of the European eel collected at 23 dpi and all Japanese eel samples collected at 23 dpi were shock-frozen in liquid nitrogen and stored at -80 °C. Storage condition was included as a factor in the relevant model (see below). RNA was extracted with TRIzol (Life Technologies, Darmstadt, Germany) following the manufacturer's recommendations for fatty tissue with slight modifications. A TissueLyzer II (Eppendorf) was used to homogenize tissue in 850 µl TRIzol. After centrifugation, another 150 µl TRIzol and 200 µl chloroform were added to the supernatant. RNA was precipitated with 500 µl isopropanol and washed with 1 ml 75% ethanol. It was resuspended in 50 µl DEPC-water (Life Technologies, Darmstadt, Germany) and incubated on a heat block at 50 °C for 2 min. Concentrations were measured with a 2100 Bioanalyzer (Agilent Technologies, Santa Clara, USA). Samples were diluted to 40 ng/µl in DEPC-water, precipitated with 3 M sodium acetate and 100% ethanol. Library preparation and paired-end sequencing (100 bp) was performed at Macrogen (Seoul, Korea) on an Illumina HiSeq4000. The number of raw reads per sample ranged from 14.6M to 30.8M (Dryad Repository).

Data analysis

Differences in length and weight between the two eel species were assessed using Wilcoxon rank-sum tests in R v.3.3.2 (R Core Team, 2016). Wilcoxon rank-sum tests were also used to determine differences in infection intensities, i.e. number of larvae in the swim bladder, between the two species within sampling days (3, 23 dpi).

Reads were *de novo* assembled into one transcriptome per species with Trinity v2.3.1 pre-release (Grabherr et al., 2011, Haas et al., 2013) using Bowtie v.1.1.2 (Langmead et al., 2009). For the European eel, raw reads from the head kidney samples from a previous study (Bracamonte et al., 2019; NCBI BioProject accession PRJNA419718) and from this study were combined. Reads from the previous study had a Phred score > 30. All reads from this study had a Phred score > 20. For the European eel, 95.95% - 96.62% of the reads per sample had a Phred score > 30. For the Japanese eel, reads with a Phred score > 30 ranged between 94.93% and 96.81% per sample. For both species, Trinity was run with default parameters, including per sample and overall in silico normalization and quality trimming using the trimmomatic option. Assembly quality and statistics were calculated using the provided Trinity scripts and Bowtie2 v2.2.9 (Langmead & Salzberg, 2012). Orthologous genes in the Japanese eel and the European eel were identified with OrthoFinder v1.1.4 using default parameters (Emms & Kelly, 2015).

Annotations for both transcriptomes were derived from blastx and blastp searches against the UniProtKB/Swiss-Prot (www.uniprot.org) and the RefSeq (www.ncbi.nlm.nih.gov/refseq/) databases. The E-value cut-off was set to 0.001. Conserved domains were identified by searching the Pfam database with HMMER v3.2.1 (hmmer.org). Annotations obtained from RefSeq were examined for their taxonomic composition and contigs that best matched bacterial sequences were removed from the transcriptomes. For the remaining contigs, GO assignments were retrieved with Trinotate v3.2.0 from annotations obtained from Swiss-Prot and Pfam. Differentially expressed genes without annotation were blasted against the nr database of NCBI (www.ncbi.nlm.nih.gov).

Gene expression was analysed separately for each species and sampling day with DESeq2 (Love et al., 2014). We could not analyse the two sampling days in a single model because sample processing differed between sampling days (see above). The need to control for the potentially large technical variation introduced by that (Leek et al., 2010) resulted in sample processing being confounded with the four treatment factors (control/infected at 3 dpi, control/infected at 23 dpi) and such models cannot be fit in DESeq2. Thus, we analysed the two sampling days in separate models. Gene-level abundance estimates were calculated using RSEM v1.3.0 (Li & Dewey, 2011). The read alignment rate for each sample ranged from 64.2% to 75.6% (Dryad Repository). Abundance estimates were modelled using generalized linear models of the negative binomial family with a logarithmic link using DESeq2 v1.14.0 (Love et al., 2014) after removing contigs with low coverage (mean coverage < 10; Todd et al., 2016). Hereafter, we refer to contigs that were maintained for differential gene expression analyses as genes. For the Japanese eel, treatment (infected, control) was included as a factor in the model for 3 dpi and in the model for 23 dpi. For the European eel, the model for 3 dpi included treatment as well as sequencing batch, the latter to control for the fact that sequencing was performed on different plates (see Bracamonte et al., 2019). The model for 23 dpi for the European eel included storage condition and treatment to control for the two sample storage conditions (see above). For all analyses, dispersion parameters were estimated with a local fit. Empirical Bayes shrinkage was applied to both dispersions and logarithmic fold changes. Genes for which expression between treatments differed by a $\log_2$ fold change ≥ 1 with an adjusted p-value < 0.05 were considered to be differentially expressed genes (DEG). P-value adjustment followed the Benjamini-Hochberg procedure as implemented in DESeq2 after independent filtering using the mean normalized count for each gene across all samples.

At 3 dpi, Japanese eel control samples were separated into two distinct clusters based on DEG (Fig. II.2). If the difference in (rlog) expression between treatment and either control cluster was smaller than the difference between control clusters, that gene was not considered to be differentially expressed (see below).

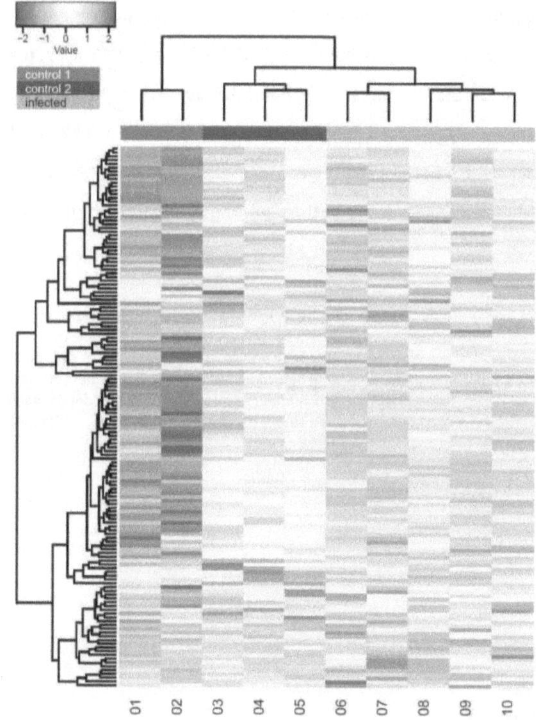

Figure II.2 Heatmap of differentially expressed genes in the Japanese eels (*Anguilla japonica*) at 3 days post-infection (dpi) prior to correction for the control group. Each column represents one sample and each row corresponds to a differentially expressed gene. Control samples form two distinct clusters (light and dark violet bars). Infected samples are indicated by a green bar. Red indicates reduced expression, blue indicates increased expression. The values are gene-level z-scores. The dendrogram is computed using Euclidean distances and clustered by the means.

DEG were used to estimate Gene Ontology (GO) enrichment with GOstats v2.48.0 (Falcon & Gentleman, 2007). GO assignments obtained with Trinotate were used as a reference. Enrichment analysis was restricted to the domain "biological processes". Conditional hypergeometric tests were performed with a p-value cutoff of 0.01. We only calculated overrepresentation of GO terms. Overrepresented GO terms were summarized with the web application CateGOrizer (Hu et al., 2008) using GO classifications available from CateGOrizer, but excluding the three general terms "metabolism", "immunology, immune response", and "response to stress". First the GO classification "Immune system gene classes" was used on all overrepresented GO terms, then "GO_Slim2" was used for GO terms that could not be summarized by immune classes, lastly, the three general terms were used on GO terms that could not be summarized by the two GO classification lists.

Results

Experimental infection

Japanese eels were significantly larger (mean ± SD 54.9 ± 7.8 cm) than European eels (41.9 ± 3.0 cm; Wilcoxon rank-sum test, W = 9, p < 0.001) and were heavier (216.8 ± 113.3 g) than European eels (113.7 ± 20.4 g; W = 9, p < 0.001). At 3 dpi, the infection intensity with *A. crassus* in swim bladders was greater in European eels (mean ± SD 2.4 ± 0.9) than in Japanese eels (0.6 ± 0.9; Fig. II.3; W = 23, p = 0.03). Only L_3 were recovered at 3 dpi. At 23 dpi, the mean infection intensity was higher in both species (Fig. II.3) but it did not differ significantly between species (Fig. II.3; W = 20.5, p = 0.12; 5.6 ± 3.4 for Japanese eels, 9.8 ± 3.4 for European eels). Both L_3 and L_4 were recovered from both eel species at 23 dpi. No adult or dead *A. crassus* were recovered. None of the control individuals were infected at any stage.

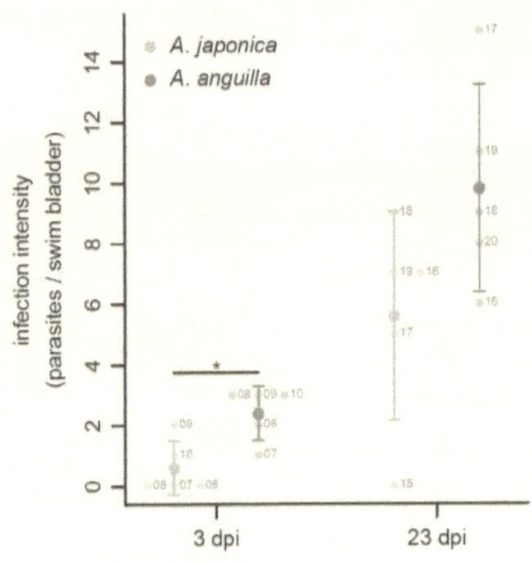

Figure II.3 Mean infection intensities (±SE) for Japanese eels (*Anguilla japonica*, yellow) and European eels (*A. anguilla*, cyan) at 3 days post-infection (dpi) and 23 dpi. Light coloured dots give number of parasites/swim bladder for each individual, * denotes significant difference between species.

Differential gene expression analysis

For the Japanese eel, 21,748,029 reads were assembled into 255,431 contigs and 347,581 isoforms with a mean coverage of 11.25. The average length of the isoforms was 556 bp and the N50 was 763 bp (Table II.1). For the European eel, 45,528,485 reads were assembled into 508,838 contigs and 693,979 isoforms. The mean coverage was 10.76. The N50 was 910 bp and the average isoform length was 610 bp (Table II.1). RefSeq annotations revealed that ~6% of the Japanese eel contigs and ~1% of the European eel contigs correspond to bacterial contamination and they were removed from further analyses.

In Japanese eels, there were 64 DEG in infected eels at 3 dpi compared to control individuals (Table II.2), with $\log_2$ fold changes ranging from -5.19 to 6.79 (Fig. II.4a). This reduced to 23 DEG at 23

Table II.1 Assembly and annotation statistics. Data for each sample are available from the Dryad Repository.

	Japanese eel		European eel	
Assembled reads	21748029		45528485	
properly paired	20669880	95.04%	42944499	94.32%
improperly paired	728604	3.35%	1879425	4.13%
single read	349545	1.61%	704561	1.55%
Number of contigs	255431		508838	
bacterial contamination	14238	5.57%	5616	1.10%
GO annotation	47429	19.66%	56977	11.32%
RefSeq annotation	62358	25.85%	136334	27.09%
Number of isoforms	347581		693979	
Mean coverage	11.25		10.76	
N50 (bp)	763		910	
Mean length (bp)	556		610	
GC content	46.53%		46.34%	

Mean coverage, N50, and mean length refer to isoforms. Percentage of GO and RefSeq annotations refer to the clean number of contigs after removal of bacterial contamination. bp, base pairs.

dpi (Table II.2), with log_2 fold changes ranging from -4.47 to 3.35 (Fig. II.4b). Only one gene was differentially expressed at both time points, with increased expression at 3 dpi and decreased expression at 23 dpi. Unfortunately, it was not annotated. There were considerably more DEG in infected European eels, with 342 DEG in infected eels at 3 dpi compared to control individuals (Table II.2). Log_2 fold changes ranged from -4.15 to 4.65 (Fig. II.4c). At 23 dpi this reduced to 53 DEG (Table II.2). Log_2 fold changes ranged from -7.61 to 6.74 (Fig. II.4d). Similarly to the Japanese eel, one unannotated gene was differentially expressed at both 3 dpi and 23 dpi, although its transcript abundance was reduced at both time points. Among the differentially expressed genes of both species and time points, 29 genes were assigned to 27 orthologous groups (Table II.3, Fig. II.4). Log_2 fold changes for the majority of orthologs were <1. Only one ortholog, a tripartite motif-containing protein, was differentially expressed in both species. Its transcript abundance was reduced in the Japanese eel but elevated in the European eel (Table II.3).

Figure II.4 (next page) Heatmaps of significantly differentially expressed genes in the Japanese eel (*Anguilla japonica*) at (a) 3 days post-infection (dpi) and (b) 23 dpi and in the European eel (*A. anguilla*) at (c) 23 dpi and (d) 3 dpi. When multiple orthologs were mapped to a differentially expressed gene in the other species, these orthologs were combined and are presented as a single row in the heatmaps. Red indicates reduced expression, blue indicates increased expression. Violet bars = control samples, green bars = infected samples. Cyan squares indicate orthologs that were differentially expressed in the Japanese eel, pink squares indicate orthologs that were differentially expressed in the European eel. For the European eel, A and B are the sequencing batches, tissue storage condition is indicated by R (RNA*later*) and N (liquid nitrogen). Gene-level z-score values are presented. Dendrograms are computed based on Euclidean distances and clustered by the means.

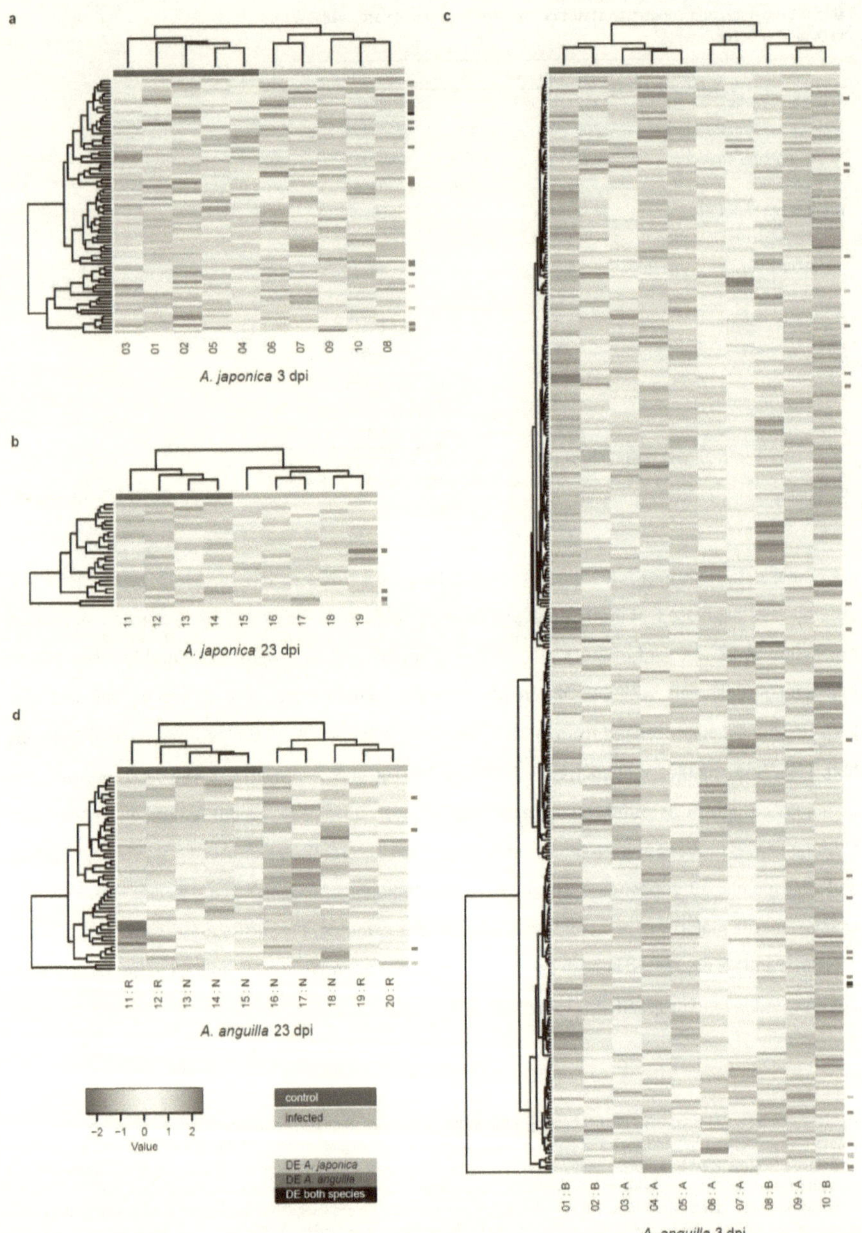

a
A. japonica 3 dpi
b
A. japonica 23 dpi
c
d
A. anguilla 23 dpi
A. anguilla 3 dpi
Value
control
infected
DE A. japonica
DE A. anguilla
DE both species

Table II.2 Number of differentially expressed genes (DEG), DEG with annotations, and overrepresented Gene Ontology (GO) terms.

Species	Dpi	Up-regulated			Down-regulated		
		total DEG	annotated	GO terms	total DEG	annotated	GO terms
Japanese eel	3	39	11	6	25	17	33
	23	11	2	0	12	1	6
European eel	3	233	139	134	209	49	41
	23	33	19	50	20	8	17

Dpi, days post-infection.

Functional analysis

For the Japanese eel, 44% of DEG at 3 dpi and 13% of DEG at 23 dpi retrieved UniProt or RefSeq annotations. This resulted in 39 overrepresented GO terms at 3 dpi and 6 overrepresented GO terms at 23 dpi (Table II.2). Most GO terms were down-regulated and associated with metabolic and cellular GO_slim2 categories (Fig. II.5a, b). At 3 dpi, processes related to an immune response were overrepresented among the down-regulated genes (Dryad Repository). These included antimicrobial response and negative regulation of TLR signalling. Transcript abundance of an immunoglobulin (Ig) lambda-like polypeptide was elevated and the abundance of *cd40* transcripts, which leads to Ig secretion via activation of B cells by T cell co-stimulation, was reduced (Dryad Repository). The abundance of erythropoietin receptor transcripts was elevated. At 23 dpi, the overrepresented GO terms for down-regulated genes were related to muscle contraction. For the up-regulated genes there were no overrepresented GO terms (Table II.2, Dryad Repository). The full lists of overrepresented GO terms and DEG can be found on the Dryad Repository.

For the European eel, annotations were retrieved for 55% of DEG at 3 dpi and 51% of DEG at 23 dpi. There were 175 overrepresented GO terms at 3 dpi and 67 overrepresented GO terms at 23 dpi (Table II.2). Unlike for the Japanese eel, GO terms were mostly up-regulated at both time points. Categories of immune system gene classes and GO_slim2 assigned them to a more diverse set of categories than the GO terms of the Japanese eel. Notably, more immune system categories were assigned for the European eel (Fig. II.5c, d). At 3 dpi, GO terms associated with an immune response were overrepresented in both up-and down-regulated genes (Dryad Repository). The regulation of cytokines, phagocytosis, and B cell activation were overrepresented in the down-regulated genes while terms related to arachidonic acid and icosanoid secretion as well as complement activation, were overrepresented in the up-regulated genes. Furthermore, transcript abundance of genes associated with wound healing was elevated. The abundance of major histocompatibility complex (MHC) class IIA and T cell receptor (TCR) α transcripts, both of which are essential for an adaptive immune response, was reduced at 3 dpi, as was the transcript abundance of six Ig chains (Dryad Repository). Genes involved in cell adhesion and motility were overrepresented among the up-regulated genes, as well as

Table II.3 Log2 fold changes of orthologous genes differentially expressed in either Japanese eels or European eels. If multiple orthologous genes were retrieved for one differentially expressed gene, log2 fold changes are given separately for each of them. Genes are sorted according to the clustering on the heatmap (Fig. II.4).

Species	UniProt or RefSeq annotation	Base mean	Log2 FC	Adj. p-value	Dpi	Ortho
Japanese eel	Complement C3	22.855	-1.722	0.245	3	g1
	Sulfate transporter	**21.866**	**-2.533**	**0.049**	3	g2
	Transmembrane protein 82	8.690	-0.933	0.474	3	g3
	Secretory phospholipase A2 receptor	8.302	-1.443	0.200	3	g4
	BTB/POZ domain-containing protein 17	9.973	-0.913	0.655	3	g5
	Collagen alpha-5(IV) chain	14.480	0.042	0.981	3	g6
	Receptor tyrosine-protein kinase erbB-3	10.106	0.372	0.785	3	g7
	Ras-related protein Rab	32.712	-1.515	NA	3	g8
	Tripartite motif-containing protein	**49.420**	**-2.497**	**0.012**	3	g9
	Ceramide synthase 2	7.987	-2.748	0.123	3	g10[a]
	Ceramide synthase 2	10.409	-0.200	0.888	3	g10[a]
	Platelet-derived growth factor C	31.101	0.267	0.858	3	g11
	Putative ribonuclease H protein At1g65750	26.217	-0.724	0.789	3	g12
	Neural cell adhesion molecule 1	15.284	-2.222	0.237	3	g13
	SLAM family member 5	12.054	-2.247	NA	3	g14
		16.941	-2.297	0.288	3	g15
	uncharacterized protein LOC106600263, partial	218.131	1.911	0.419	3	g16
	Wilms tumor protein homolog	93.910	0.126	0.850	3	g17
	Estrogen receptor beta	46.955	0.154	0.901	3	g18
	Transmembrane protein 150B	**274.372**	**-4.660**	**0.000**	3	g19
	Erythropoietin receptor	**1631.368**	**4.064**	**0.010**	3	g20
	Ig lambda-1 chain V region S43	1099.355	-0.074	0.963	3	g21
	Olfactomedin-4	**671.199**	**-1.665**	**0.003**	3	g22
	Cytochrome P450	76.495	-2.232	0.219	3	g23[a]
	Cytochrome P450	116.302	0.369	0.752	3	g23[a]
European eel	Wilms tumor protein homolog	**40.787**	**1.589**	**0.016**	3	g17
	Receptor tyrosine-protein kinase erbB-3	**15.722**	**1.607**	**0.028**	3	g7[b]
	uncharacterized protein LOC106600263, partial	**35.068**	**2.528**	**0.017**	3	g16
	Ceramide synthase 2	**26.228**	**1.969**	**0.036**	3	g10
	Tripartite motif-containing protein	34.275	0.491	0.949	3	g9[b]
	Tripartite motif-containing protein	73.534	-0.527	0.896	3	g9[b]
	Tripartite motif-containing protein	62.575	-0.279	0.963	3	g9[b]
	Tripartite motif-containing protein	100.232	-0.363	0.898	3	g9[b]
	Tripartite motif-containing protein	22.121	-0.540	0.969	3	g9[b]
	Transmembrane protein 82	**16.591**	**2.539**	**0.004**	3	g3
	Receptor tyrosine-protein kinase erbB-3	**13.040**	**2.184**	**0.003**	3	g7[b]
	Ras-related protein Rab	**18.986**	**2.448**	**0.011**	3	g8
	Putative ribonuclease H protein At1g65750	**46.092**	**3.034**	**0.024**	3	g12
	Neural cell adhesion molecule 1	**69.045**	**-2.592**	**0.025**	3	g13
	SLAM family member 5	**12.386**	**-1.902**	**0.003**	3	g14
	Estrogen receptor beta	**242.160**	**2.552**	**0.020**	3	g18
	Collagen alpha-5(IV) chain	**88.892**	**1.837**	**0.025**	3	g6
	Complement C3	**522.554**	**2.495**	**0.036**	3	g1
		239.109	**1.979**	**0.045**	3	g15
	Secretory phospholipase A2 receptor	**48.205**	**1.514**	**0.018**	3	g4
	Tripartite motif-containing protein	**59.652**	**1.609**	**0.014**	3	g9[b]
	Platelet-derived growth factor C	**49.289**	**1.657**	**0.013**	3	g11
	Erythropoietin receptor	69.219	-0.708	0.847	3	g20
	BTB/POZ domain-containing protein 17	**36.409**	**-2.534**	**0.000**	3	g5
	Ig lambda-1 chain V region S43	**153.169**	**-3.156**	**0.000**	3	g21
	Olfactomedin-4	765.320	-0.495	0.809	3	g22
	Transmembrane protein 150B	72.345	3.167	NA	3	g19
	Cytochrome P450	**816.051**	**1.866**	**0.029**	3	g23
	Sulfate transporter	586.823	0.280	0.849	3	g2

		Base mean	Log2 FC	Adj. p-value	Dpi	Ortho
Japanese eel	Apolipoprotein A-I-1	8.782	0.491	NA	23	g24
	LPS-induced TNF-alpha factor homolog	34.972	0.316	1.000	23	g25
	Probable G-protein coupled receptor 34	112.783	-0.204	1.000	23	g26[a]
	Probable G-protein coupled receptor 34	37.250	0.346	1.000	23	g26[a]
	40S ribosomal protein S27-like	**24422.925**	**2.427**	**0.026**	23	g27
European eel	Apolipoprotein A-I-1	**60.270**	**5.024**	**0.001**	23	g24
	LPS-induced TNF-alpha factor homolog	**9.189**	**4.920**	**0.039**	23	g25
	Probable G-protein coupled receptor 34	**76.144**	**-1.841**	**0.000**	23	g26
	40S ribosomal protein S27	19585.355	0.118	0.962	23	g27

Base mean, mean expression across all samples, Log2 FC, log2 fold expression change between infected and control samples, Adj. p-value, Benjamini-Hochberg-adjusted p-value calculated by DESeq2, Dpi, days post-infection, Ortho, orthologous groups, significantly different log2 fold changes are given in bold.
[a]multiple orthologous genes in the Japanese eel
[b]multiple orthologous genes in the European eel

genes involved in renal development and function (Dryad Repository). Transcript abundance of oxygen-dependent coprophorphyrinogen-III oxidase (*cpox*), an enzyme involved in heme biosynthesis, was reduced. The abundance of N-acetyl-D-glucosamine kinase (*nagk*) transcripts, involved in amino sugar metabolism, was also reduced (Dryad Repository). At 23 dpi, transcript abundance of genes related to an adaptive immune response (e.g., MHC IIB) and inflammation (Dryad Repository) was elevated. GO terms related to cellular respiration were overrepresented among down-regulated genes (Dryad Repository). More specifically, mitochondrial cytochrome c oxidase subunits I and III (*cox1* and *cox3*) and cytochrome b (*cytb*) all had reduced transcript abundances (Dryad Repository). The full lists of overrepresented GO terms and DEG for the European eel are available from the Dryad Repository.

Discussion

Anguillicola crassus is a parasitic swim bladder nematode native to the Japanese eel (*Anguilla japonica*) that was first detected in the European eel (*A. anguilla*) population approximately 35 years ago (Kirk, 2003). Infection intensities measured in natural populations (Audenaert et al., 2003, Gérard et al., 2013, Heitlinger et al., 2009, Knopf, 2006, Münderle et al., 2006) and in experimental individuals several weeks after an infection was established (Knopf & Mahnke, 2004, Knopf & Lucius, 2008) indicate greater susceptibility by the novel host, the European eel. The population of the European eel has undergone catastrophic declines (Bornarel et al., 2017, Diekmann et al., 2019, ICES, 2018), and *A. crassus* infections have been implicated (Drouineau et al., 2018, Sures & Knopf, 2004). We found similar infection intensities in the two eel species 23 days after experimental infection, at which time larval parasites had finished migrating to the swim bladder, and we did not find any dead larvae. Although our sample size was small, our findings support the observations at 25 dpi by Weclawski et al. (2013) and indicate that the different abilities of the two eel species to clear an infection do not manifest early after infection, and may only become apparent after more advanced developmental stages of the parasite are present. The low number of parasites in the swim bladders we observed at 3 dpi in both eel species indicates that a considerable proportion of larvae was still migrating towards the swim bladder.

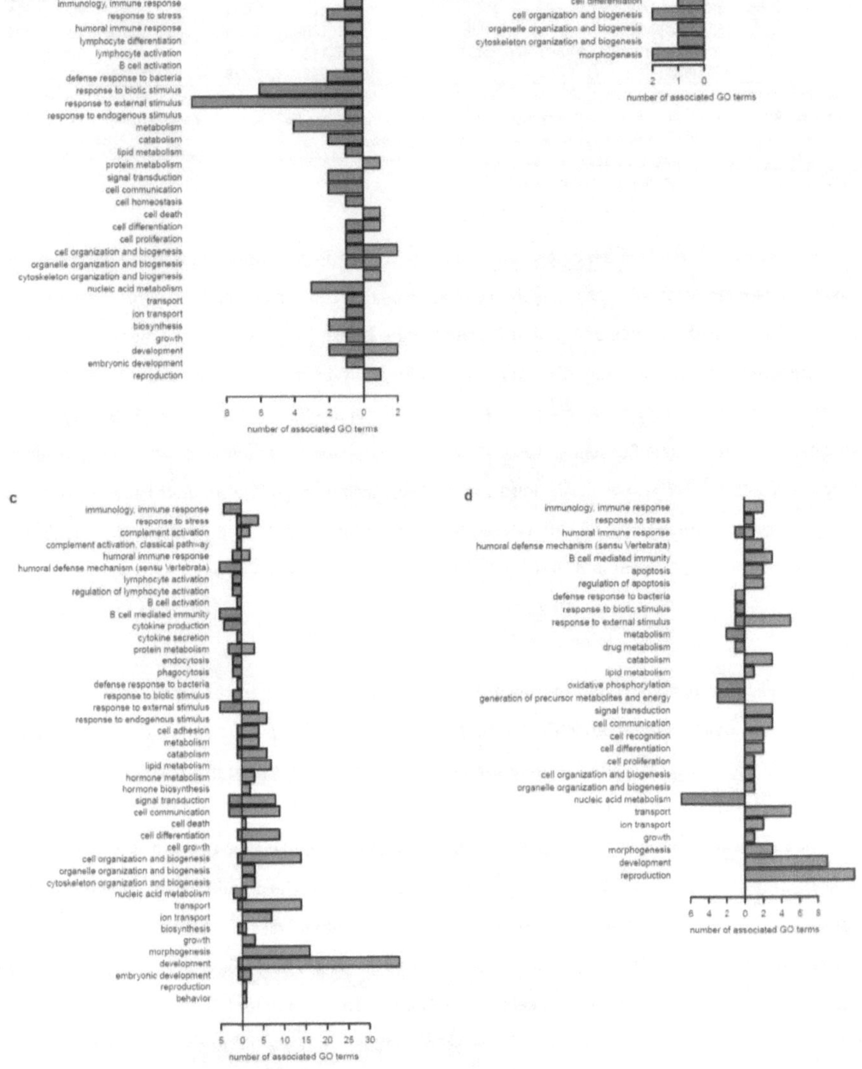

Figure II.5 Overrepresented Gene Ontology terms categorized into immune classes and GO_slim2 in (a) the Japanese eel (*Anguilla japonica*) at 3 days post-infection (dpi) and (b) 23 dpi and the European eel (*A. anguilla*) at (c) 3 dpi and (d) 23 dpi. Blue bars to the right indicate enrichment in up-regulated genes, red bars to the left indicate enrichment in down-regulated genes of infected individuals.

We found almost 5 times as many DEG in infected European eels than in infected Japanese eels. The difference in expression between species was most pronounced during the parasite migration phase (3 dpi) but remained large after parasites were established in the swim bladder wall (23 dpi). In both species, more genes were differentially expressed during the larval migration than after establishment, although this temporal difference was much more pronounced for the European eel. This could indicate that the tissue damage caused by the migrating parasites is more problematic for both eel species than the presence of larvae in the swim bladder wall. We cannot fully exclude the possibility that, although all individuals were handled identically, the infection procedure affected experimentally and sham-infected individuals differently and that, at 3 dpi, recovery from stress of handling contributed to the overall greater number of DEG.

In the Japanese eel, immune system processes were only weakly affected, and only at 3 dpi. Those processes that were affected did not indicate that the Japanese eel attempted to clear the infection. Reduced abundance of *cd40* transcripts might indicate non-responsiveness of the immune system during *A. crassus* migration to the swim bladder. CD40 is essential for initiating a T cell-dependent adaptive immune response. By providing a co-stimulatory signal, it leads to activation, proliferation, and differentiation of lymphocytes and induces maturation of dendritic cells during interaction with T cells. The absence of co-stimulation induces immunological tolerance rather than immunity (Elgueta et al., 2009, Quezada et al., 2004). At 23 dpi there was no evidence for an immune response. However, only a small proportion of DEG could be annotated. While this may indicate that some species-specific genes could be involved in the response, we do not expect that it would cause an immune response to remain undetected because the fish immune system is very similar to the well-described mammalian immune system (Buchmann, 2012, Alvarez-Pellitero, 2008).

Similarly to the infection intensities we and Weclawski et al. (2013) observed, this lack of response suggests that an early and efficient immune response does not contribute to lower susceptibility observed in the Japanese eel. An important consideration is that most studies have only inferred the immune response based on infection intensities and proportions of dead *A. crassus* (Knopf & Mahnke, 2004, Weclawski et al., 2013). Other studies have assessed a response based on antibodies in the presence of adult *A. crassus*, but antibodies may not be crucial for reducing infection intensities in this system (Knopf & Lucius, 2008, Nielsen, 1999). Large numbers of dead *A. crassus* larvae in the Japanese eel were reported after adult *A. crassus* had appeared (Heitlinger et al., 2009, Weclawski et al., 2013). Taken together, there is no evidence that would suggest that the Japanese eel mounts an immune response in the early stages of first infection, and a response to any parasitic stage, including larvae that have halted development (Knopf & Mahnke, 2004), may only occur after adult stages are present in the swim bladder. However, infection with larvae that were exposed to a sublethal dose of radiation which impaired their development to adulthood led to encapsulation several weeks post-infection (Knopf & Lucius, 2008).

Infected European eels underwent gene expression changes in not only more genes, but in a wider range of processes that included immune response, renal function, and energy generation. In the Japanese eel, none of the differentially expressed genes were associated with renal function or energy budget, although erythropoiesis (production of red blood cells) may have been initiated at 3 dpi. This adds to previous findings in amphibians (Ellison et al., 2015, Eskew et al., 2018, Poorten & Rosenblum, 2016), mammals (Davy et al., 2017, Field et al., 2015), and even honeybees (Zhang et al., 2010) and indicates that a stronger response to an invasive parasite in susceptible host species than in resistant host species may be a general pattern in the response to parasites as diverse as fungi, mites, and helminths. One important consideration is that this and other studies have focused on one or few time points during infection. Resistant hosts may produce a strong response at an as yet unstudied time point during infections that leads to reduction and clearance of parasites. In contrast, susceptible hosts may respond inadequately throughout the infection.

Immune genes that were differentially expressed in the European eel indicated that eels produced an inflammatory response to *A. crassus* throughout the experiment. This did not appear to induce an adaptive immune response early during the infection because transcript abundance of genes encoding MHC IIA, TCR α, and immunoglobulin chains was reduced at 3 dpi. The abundance of *Mhc IIB* transcripts was elevated following the establishment of parasite larvae in the swim bladder wall, which may indicate that the adaptive immune response was activated at 23 dpi. The activation of immune responses was not reflected in lower infection intensities in the European eel compared to the Japanese eel, suggesting that the early response of the European eel is ineffective. Considering that the Japanese eel did not initiate an immune response this early after infection, the timing of the response by the European eel may also be inappropriate and contribute to its inefficiency. A previous study found considerably more immune genes to be differentially expressed in the swim bladder of naturally infected European eels (Schneebauer et al., 2017). There are several reasons that might account for the differences between studies. First, we experimentally infected the eels once with a moderate number of *A. crassus* larvae while Schneebauer et al. (2017) used naturally infected wild-caught eels that contained parasites at all stages of the life cycle including adults. The European eel was previously shown to produce antibodies against adult *A. crassus* (Knopf & Lucius, 2008, Nielsen, 1999), thus the presence of adult parasites may enhance the immune response, as discussed above for the Japanese eel. Second, the response to an infection differs among organs (Bracamonte et al., 2019, Huang et al., 2016, Robledo et al., 2014). If an immune response is induced locally but not systemically, it would only be detectable in the infected organ. And third, infected and control yellow eels of Schneebauer et al. (2017) originated from different geographic locations and were likely exposed to differing parasite communities (Gérard et al., 2013, Kennedy, 2001). The immune response might therefore be influenced by the prevailing parasite community rather than the presence of a single parasite species (Huang et al., 2016, Stutz et al., 2015). Our eels were kept within a recirculation system presumably free of parasites. We did not screen for additional parasites, but any other parasites present would have been homogenized throughout the system and therefore equally exposed all eels in the experiment.

Similarly to susceptible amphibians and mammals, the European eel modified the expression of genes involved in energy generation, suggesting that an infection affects maintenance of homeostasis. Transcript abundance of *cpox* which encodes an enzyme of the heme biosynthesis pathway (Layer et al., 2010) was reduced at 3 dpi. This contrasts with our previous finding of increased ferrochelatase expression in the spleen of infected European eels (Bracamonte et al., 2019). Heme is an essential part of hemoproteins such as hemoglobin and cytochromes. The expression of hemoglobin, the oxygen carrier of red blood cells, is affected by *A. crassus* infections, though after the appearance of adult parasites (Fazio et al., 2009). Here we found reduced transcript abundance of several cytochromes of the respiratory chain at 23 dpi. This may be a consequence of earlier (3 dpi) reduction of heme biosynthesis. Cell respiration provides energy and its reduction in infected European eels could lead to poor performance during energetically costly activities (Palstra et al., 2007).

Immune responses carry energetic and physiological costs, and excessive and inappropriate regulation and timing of the immune response can cause tissue damage, can interact with maintenance of homeostasis or reproduction, and can lead to fitness loss and even host death even if the infection is cleared (Graham et al., 2005, Lochmiller & Deerenberg, 2000, Sheldon & Verhulst, 1996). Reducing an immune response and its associated costs can therefore be beneficial for the organism. Stated another way, tolerating infections, i.e. mitigating the negative effect of an infection on host health (Raberg et al., 2009), may be more beneficial than clearing them if the self-inflicted damage caused by an immune response outweighs the costs of bearing an infection (Graham et al., 2005, Read et al., 2008). Considering that we infected eels once with a moderate number of *A. crassus* and we sampled before the blood-feeding parasitic stages appeared, destroying and degrading nematode tissue might be too costly at this stage of infection. The inappropriate immune response that we observed in the European eel compared to the Japanese eel may have promoted the disruption of the non-immune processes. This may then translate into higher health and fitness costs despite the absence of differing infection intensities at this early stage of infection. However, blood-feeding adult *A. crassus* and continuous exposure, as seen in wild eels (Heitlinger et al., 2009), might cause more damage and trigger a noticeable protective immune response in the Japanese eel.

Limitations of the study

Our study provides a robust experimental comparison of both species under identical conditions, but some limitations of the study hinder a more complete understanding of the response. First, the number of DEG that were annotated was always lower than 50%, albeit similar for the European eel at both time points and for the Japanese eel at 3 dpi. The value was much lower for the Japanese eel at 23 dpi. This is a common problem in RNA-seq studies and fundamentally hampers our understanding in non-model organisms (Pavey et al., 2012). Eels are fish, but evolutionarily distant from model fish species with more completely annotated genomes. Second, the possibility that the European eels were silvering, i.e. preparing for long-distance oceanic migration, makes our species comparison more conservative, based on the observation that silvering leads to an overall reduction in the number of DEG compared to yellow eels (Schneebauer et al., 2017). A consequence is that our comparison may underestimate the differences between species. Although we did not measure eye diameter, we observed that European eel but not Japanese eel individuals had enlarged eyes, which is a sign of silvering (Tesch, 2003). Silvering involves morphological and physiological modifications (Tesch, 2003, Durif et al., 2005). These modifications are energetically costly and might reduce resource allocation to other functions, including the immune system. Finally, hosts and parasites undergo coevolution. Ongoing differentiation and adaptation of *A. crassus* populations (Heitlinger et al., 2014, Weclawski et al., 2013) may limit the suitability of using infections with the European parasite population as null model for the Japanese eel responses, although fitness parameters and overall gene expression of adults do not differ between Asian and European parasites (Heitlinger et al., 2014, Weclawski et al., 2013, Weclawski et al., 2014). As with silvering (above), using European parasites means that the differences between host species that we observed were conservative, and that the European response may be even more pronounced than the Japanese, if the Japanese eel in our experiment was potentially less adapted to the parasite than the European eel was.

Conclusion

The European eel is undergoing catastrophic population declines and infection with *Anguillicola crassus* could be one of the contributing factors (Drouineau et al., 2018). Based on comparison of the number and diversity of differentially expressed genes with the Japanese eel, we conclude that the European eel has not adapted to *A. crassus*. However, it may be that responses to infection soon after the parasite's introduction were much stronger than what we observed here, and that some degree of adaptation has taken place in the 35 years since introduction (approx. 3-5 eel generations). The comparison between the two species further indicates that preventing disruption of metabolic and physiological processes is imperative for reducing susceptibility. In contrast, producing an immune response immediately after first contracting the parasite may not provide sufficient benefits. While the impact of parasites on host physiology or fitness may be a better estimate of the potential threat than immune response or parasite load (Viney et al., 2005), determining the timing and associated costs of the immune response in the native host can help to clarify the optimal defence strategy and provide a baseline for identifying possible adaptation by additional hosts.

Data Accessibility

Raw RNA-seq reads were deposited in the BioProjects PRJNA419718 (European eel 3 dpi), PRJNA546508 (European eel 23 dpi), and PRJNA546510 (Japanese eel 3 and 23 dpi). Supplementary data are available from the Dryad Repository

Encapsulation of *Anguillicola crassus* reduces the number of adult parasite stages in the European eel (*Anguilla anguilla*)

Seraina E. Bracamonte[1,2,3,*], Klaus Knopf[1,3], Michael T. Monaghan[1,2,4]

[1]Leibniz-Institute of Freshwater Ecology and Inland Fisheries, Müggelseedamm 301/310, 12587 Berlin, Germany
[2]Berlin Center for Genomics in Biodiversity Research, Königin-Luise-Strasse 6-8, 14195 Berlin, Germany
[3]Faculty of Life Sciences, Humboldt-Universität zu Berlin, Invalidenstrasse 42, 10115 Berlin, Germany
[4]Institut für Biologie, Freie Universität Berlin, Königin-Luise-Strasse 1-3, 14195 Berlin, Germany

Unpublished

Introduction

Invasion by non-native parasites can affect the viability of novel hosts (Peeler et al., 2011, Daszak et al., 2000), posing strong selection pressure on the host to adapt (Penczykowski et al., 2011). Adaptation has been observed in response to novel parasites. A Hawaiian honeycreeper species shows signs of increased tolerance to avian malaria after severe population declines following the parasite's introduction (Atkinson et al., 2013). Both increased tolerance and increased resistance were reported in blue mussels after the introduction of a parasitic copepod (Feis et al., 2016). Increased resistance was also observed in a rainbow trout population in response to an invasive myxozoan parasite (Miller & Vincent, 2008).

Anguillicola crassus Kuwahara, Niimi & Hagaki is a parasitic swim bladder nematode that is invasive in the European eel (*Anguilla anguilla*, L.). It was first detected in European fresh waters in 1982 and it has rapidly spread across the entire range of its new host (Kirk, 2003). Infections with the parasite were suggested to hamper the trans-oceanic spawning migration and reproduction of the European eel (Palstra et al., 2007, Pelster, 2015). Thus, it may contribute to the dramatic population decline observed in the last decades (Bornarel et al., 2017, Drouineau et al., 2018, Diekmann et al., 2019) and the European eel's status as critically endangered (Jacoby et al., 2015). Consequently, European eel individuals that can cope with *A. crassus* should have an advantage over non-responders.

The Japanese eel (*A. japonica* Temminck & Schlegel), the parasite's native host, frequently encapsulates *A. crassus* in natural infections (Münderle et al., 2006, Heitlinger et al., 2009). Encapsulation is considered the main reason for lower infection intensities and lower impact on fitness in the Japanese eel compared to the European eel (Taraschewski, 2006). The European eel is capable of encapsulating *A. crassus* (Molnár, 1994) and the frequency has increased from 0% in 1990 to 20% in 1997 and 2000 in Flanders, Belgium (Audenaert et al., 2003). Hence, encapsulation may be a means of adaptation by the European eel.

European eels infected with *A. crassus* are susceptible to adverse environmental conditions (Molnár et al., 1991). Mortality during hypoxia increases with severity of infection (Molnár, 1993, Lefebvre et al., 2007). Infected eels consume more oxygen during activity than non-infected, thus having a higher energy demand (Palstra et al., 2007). Consistent with higher oxygen demand with infection severity, expression of the haemoglobin α gene correlated with parasite biomass in experimental infections (Fazio et al., 2009). Down-regulation of several cytochrome genes of the cell respiration pathway, however, indicate that energy provision may be compromised in experimentally infected European eels (Bracamonte et al., in press).

Populations regularly differ in their resistance to parasites which appears to be related to the degree of exposure and adaptation (Weber et al., 2017, MacColl & Chapman, 2010). Populations that are adapted to a particular parasite were shown to mount a stronger immune response when challenged with that parasite (Kalbe & Kurtz, 2006, Scharsack & Kalbe, 2014). Increased immune gene expression has been associated with higher resistance in fish (Lohman et al., 2017, Lenz et al., 2013), birds (Bonneaud et al., 2011), and mammals (Guo et al., 2016). At the same time there is evidence that the expression of non-immune genes is also differently affected in populations differing in parasite resistance (Bonneaud et al., 2011). Additionally, individuals of better adapted populations tend to grow more, have better body conditions, and higher metabolic condition when infected, suggestive of reduced metabolic and energetic costs (MacColl & Chapman, 2010, Kalbe & Kurtz, 2006, Kurze et al., 2016). Based on these observations in other species, we hypothesize that encapsulating, i.e. adapting to, *A. crassus* leads to lower infection intensities in the European eel, an increased immune response, and reduced metabolic costs.

For European eels from Lake Müggelsee, Berlin, Germany, we identified *A. crassus* infection intensity and macroparasite community composition and compared them between eels encapsulating *A. crassus* and those not encapsulating it. We further tested for temporal variation of these parameters between August and October. We used quantitative PCR to test if eels encapsulating and those not encapsulating the parasite differed in immune (*mhc II*), energy-related (*cox1*), and haematopoietic (*epor*) gene expression suggestive of an increased immune response and reduced metabolic costs. As for infection intensity and parasite community, we tested if gene expression responses showed temporal variation. Genes were selected based on differential expression in transcriptome-wide expression studies on European eels and Japanese eels experimentally infected with *A. crassus*. *Mhc IIA* and *mhc IIB* both had altered expression profiles in infected European eels (Bracamonte et al., 2019, Bracamonte et al., in press). We expected increased expression of *mhc II* genes in eels encapsulating *A. crassus*, because encapsulation may be an evolved response. *Cox1* expression was reduced in European eels experimentally infected with *A. crassus* (Bracamonte et al., in press). We hypothesized that individuals encapsulating *A. crassus* would have higher expression, because encapsulation should mitigate disruption of the energy balance. Expression of *epor* was increased in Japanese eels following *A. crassus* infections (Bracamonte et al., in press). We expected increased expression in more heavily infected European eels, especially in the presence of many blood-feeding adult parasites. Furthermore, we expected reduced expression in the individuals encapsulating *A. crassus*, because encapsulation supposedly reduces the intensity of infection.

Materials and Methods

Sampling

European eels were caught by electrofishing near Surferwiese (52.448° N 13.656° E (SF)) in Lake Müggelsee, Germany, on 8 August 2017 and on 10 and 17 October 2017. Eels were immediately decapitated, immobilized by destruction of the spinal cord, and kept on ice for transportation back to the laboratory. Dissections were carried out approximately one hour after fishing. The spleen and the head kidney were removed and stored at -20 °C in RNA*later* (Life Technologies, Darmstadt, Germany) following the manufacturer's instructions. For all individuals, weight was determined to the nearest g and total length (TL) to the nearest 0.5 cm. Relative condition factor (K_{rel}) was calculated according to Le Cren (1951) using the values available from FishBase (Froese & Pauly, 2019). *A. crassus* in the swim bladder and other parasites on the gills, in the gut, the anal fin, and the eyes were counted using a stereo microscope (7x-70x magnification). The intestinal cestodes *Bothriocephalus claviceps* and *Proteocephalus macrocephalus* were combined, because they were assumed to be biologically similar and because they could not always be distinguished during dissection. Similarly, the gill monogeneans *Pseudodactylogyrus bini* and *Pseudodactylogyrus anguillae* were not recorded separately. Cysts formed by the myxozoans *Myxobolus portucalensis* on the anal fin and *Myxidium giardi* on the gills and *Pseudodactylogyrus* spp. were categorized into abundance classes of 0, 1-5, 6-20, and >20 (Table SIII.1). The number of encapsulated *A. crassus* larvae in the swim bladder wall was recorded.

Analysis of parasite communities

All analyses were done in R v3.5.3 (R Core Team, 2019). Prevalence, mean infection intensity, and mean abundance of *A. crassus* (third and fourth stage larvae (L_3 and L_4) and adult stages), larval *A. crassus* (L_3 and L_4), and adult *A. crassus* were calculated for all eels and separately for each month. Throughout the manuscript, we will refer to L_3, L_4, and adult stages as *A. crassus* and to L_3 and L_4 as larvae or larval stages, i.e. excluding eggs and L_2. We determined the prevalence of capsules. We also calculated infection intensity and abundance of larval and adult stages only including eels that contained living *A. crassus* in the swim bladder and for which the encapsulation status (presence/absence of capsules) could be determined unambiguously. For these eels we used Wilcoxon rank-sum tests to estimate if weight and length differed between encapsulation status or sampling month. Six eels either did not contain living *A. crassus* in their swim bladders or were of uncertain encapsulation status and were excluded from further analyses (Table SIII.1).

We assessed differences in overall abundance, abundance of larval *A. crassus*, and abundance of adult *A. crassus* with generalized linear models (GLM) that included encapsulation status, sampling month, and their interaction as factors, each with a Poisson distribution with a log link function using the glm function. We performed Tukey's HSD *post hoc* tests using the multcomp v1.4-8 (Hothorn et al., 2008) package for R. We correlated the total number of *A. crassus*, the number of larval *A. crassus*, and the number of adult *A. crassus* with weight and TL using Spearman's rank correlation tests with the cor.test function. We performed these tests once for all eels and then separately for eels sampled in August and October and for eels of the non-encapsulating group (i.e. without capsules, NC) and the encapsulating group (i.e. with at least one capsule, C).

Prevalences and, if applicable, mean abundances and mean intensities for all other parasites were estimated overall and separately for August and October and for each encapsulation status (NC and C). GLM were used to test if prevalences of each parasite differed between encapsulation status and month applying a binomial distribution with a logistic link function. Similarly, GLM with a Poisson distribution with a log link function were used to test for differences of abundances. Differences in parasite community composition, excluding *A. crassus* and species with < 10% overall prevalence, were determined with an analysis of similarity (Clarke, 1993) on Bray-Curtis distances using the function anosim of the vegan package v2.4-4 (Oksanen et al., 2017) for R with 100000 permutations. Parasite communities were compared between the two months and the two encapsulation status. Species that contributed most to parasite community dissimilarities were identified with a similarity percentage analysis (simper) implemented in the vegan package. For visualization, non-metric multidimensional scaling plots were produced with the function metaMDS of the vegan package.

RNA extraction and cDNA synthesis

RNA was extracted from spleen and head kidney tissue as described in Bracamonte et al. (2019) and quantified on a NanoDrop 1000 Spectrophotometer (Thermo Scientific, Darmstadt, Germany). Remnant DNA was removed from RNA extracts with DNase I, Amplification grade (Thermo Fisher Scientific) following the manufacturer's instructions. Purified RNA was reverse-transcribed in duplicates with MMLV High Performance Reverse Transcriptase (Biozym, Hessisch Oldendorf, Germany) following the manufacturer's instructions. For quantitative real-time PCR (qPCR), duplicate reverse transcriptions were pooled.

Target genes

The selected genes had either increased expression (*mhc IIB* and *epor*) or decreased expression (*mhc IIA* and *cox1*) in infected individuals (Bracamonte et al., 2019, Bracamonte et al., in press), could be assigned to distinct physiological processes, and can be expected to be modified if encapsulation is the result of a specific and advantageous response to *A. crassus*. The major histocompatibility complex class II (MHC II) is essential for initiating an adaptive immune response against extracellular parasites which ultimately results in highly specific antibody production (Morris et al., 1994). An MHC II molecule is composed of two chains encoded by genes *mhc IIA* and *mhc IIB*. *Mhc IIB* is usually more polymorphic, providing higher antigen specificity (Reche & Reinherz, 2003, Brown et al., 1993). However, in European eels *mhc IIA* may be equally variable (Bracamonte et al., 2015). Cytochrome c oxidase subunit 1 (COX1) is a core protein of the respiratory chain which is responsible for energy generation (Hosler et al., 2006). The erythropoietin receptor (EPOR) is expressed on the progenitors of erythrocytes during their maturation and promotes their proliferation and differentiation (Elliott et al., 2014).

Primer design and qPCR

We carried out qPCR using a combination of newly designed and published primers (Table III.1). We newly designed primers for *cox1*, *epor*, and the housekeeping gene β-actin (*actb*) using sequences from two European eel transcriptome assemblies (Bracamonte et al., 2019, Bracamonte et al., in press) and other data as follows. For *cox1*, we included sequences of European eels and American eels (*A. rostrata*) available on NCBI (acc. nos. NC_006531.1 and NC_006547.2) and sequences from EeelBase (Coppe et al., 2010). For *epor*, no anguillid sequences were available in public databases, therefore we included sequences of a Japanese eel transcriptome assembly for primer design (Bracamonte et al., in press). For *actb*, we used a published reverse primer (Fazio et al., 2008a) and a new forward primer designed using European eel and Japanese eel sequences available on NCBI (acc. nos DQ286836.1, KJ021893.1, GU001950.1) and EeelBase (Coppe et al., 2010). Primers for both *mhc II* genes were modified from Bracamonte et al. (2015).

Table III.1 Oligonucleotide sequences used for PCRs and qPCRs and amplicon size.

Gene	Primer name	Sequence 5' → 3'	Amplicon size	Source
Actb	ACTBF2	GAGACCACCTTCAACTCC	196 bp	present study
	Actin R	TCCAGACGGAGTATTTGC		Fazio et al., 2008
Cox1	COX1F2	CTACTCCTCTCCCTGCCAGT	150 bp	present study
	COX1R2	GTATACTTCTGGGTGGCCGA		present study
Epor	EPORF1	ACAATGACACGGACAGGGAA	142 bp	present study
	EPORR1	CCTTCACCAATTCCCGCTTG		present study
Mhc IIA	MHCIIAE3F	GATCCTCCTCAGAGCACAATCT	250 bp	modified from Bracamonte et al., 2015
	MHCIIAE3R	TGTGCTCCACGCTGCAGGAA		modified from Bracamonte et al., 2015
Mhc IIB	MHCIIBE3F3	TTCTACCCCAGAGGAATCAAAATGAC	167 bp	Bracamonte et al. 2015
	MHCIIBE3R	TGCTCCACCTTGCAGGAGATTT		modified from Bracamonte et al., 2015

bp, base pairs

Primers were validated in regular PCRs using the Biozym Probe qPCR Kit (Biozym). PCRs were carried out in a volume of 20 µl containing 10 µl of 2x qPCR Probe Mix, 400 nM of forward and reverse primer, and 2 µl of cDNA for the genes *actb*, *cox1*, *mhc IIA*, and *mhc IIB*. PCR for *epor* contained 500 nM of each primer. For genes *actb*, *cox1*, *mhc IIA*, and *mhc IIB*, cycling conditions were as follows: initial denaturation at 95 °C for 2 min, 30 cycles of 95 °C for 5 s and 65 °C for 30 s, and a final elongation at

65 °C for 10 min. For *epor*, the number of cycles was increased to 40. A Mastercycler nexus GSX1 (Eppendorf, Hamburg, Germany) was used for all PCRs. PCR products were purified and sequenced at Macrogen Europe (Amsterdam, The Netherlands). Sequences were aligned back to those used for primer design and they were blasted against the nr protein database of NCBI for identity confirmation.

For the qPCRs, the reaction mix was identical to that used for regular PCRs (above) except that 0.0006 µl 10,000x SYBR Green I Nucleic Acid Gel Stain (Invitrogen, Darmstadt, Germany) were added. Reactions were run on a Stratagene Mx3005P qPCR system (Agilent Technologies, Waldbronn, Germany) with the cycling conditions described above, but omitting the final elongation. The number of cycles was set to 40 for all genes. QPCRs were run in duplicates and a 5-fold dilution series was added on each plate as standard curve. Ct values were determined with MxPro QPCR Software (Agilent Technologies) using default parameters.

Gene expression analysis

Relative expression of the target genes was calculated for every sample following the $\Delta\Delta$Ct method of Pfaffl (2001) using *actb* as reference gene. The 1:5 dilution of the standard curve was used as a calibrator to standardize among plates. Since tissues are known to differ in gene expression, the spleen and the head kidney were analysed separately. Analyses were performed separately for every gene in R. GLMs were used to analyse relative gene expression changes as a function of encapsulation status, sampling month, and their interaction. Tukey's HSD *post hoc* tests were performed for models with significant factors. Relative gene expression changes were further analysed with infection intensity of *A. crassus* as continuous predictor, sampling month as a factor, and their interaction. The same analyses were performed with the number of adult *A. crassus* as continuous predictor, sampling month as a factor, and their interaction separately for the NC group and the C group. For the C group, an additional model included the number of capsules as continuous predictor, sampling month as a factor, and their interaction. QPCR plate was included as a fixed factor in all models. The models for *mhc IIA* additionally included all two-way interactions with qPCR plate. For the *cox1* gene, residuals were normally distributed, thus a Gaussian distribution was used. For the genes *mhc IIA*, *mhc IIB*, and *epor*, a Gamma distribution was used with an inverse link function with few exceptions detailed below. For *epor* in the spleen, starting values of 0.5 for the intercept and 0 for the predictors were supplied to the model with capsules as continuous predictor. For *mhc IIA* in the spleen, a Gaussian distribution was used for all models including adult *A. crassus* as a predictor. For *mhc IIA* in the head kidney, a Gaussian distribution was used for the models of the C group including capsules or adult *A. crassus* as predictor. Spearman's rank correlation tests were used to assess correlation between the relative expression levels of *mhc IIA* and *mhc IIB*.

Results

Mean weight ± SE, mean TL ± SE, and mean K_{rel} ± SE were 122 ± 15 g, 40.6 ± 1.5 cm, and 0.948 ±

0.014 and they did not differ significantly between C and NC eels (Wilcoxon rank test, weight: W = 97,

p = 0.32, length: W = 94.5, p = 0.27, K_{rel}: W = 135, p = 0.67). Weight and TL were greater in August than

in October (W = 171, p = 0.029 for both weight and TL, August: 166 ± 26 g and 45.5 ± 2.6 cm, October:

100 ± 16 g and 38.0 ± 1.6 cm) but there were no differences in K_{rel} (W = 111, p = 0.87, August: 0.941 ±

0.022, October: 0.951 ± 0.018).

Table III.2 Prevalence (P, %), mean intensity ± SE (I), and mean abundance ± SE (A) of living *A. crassus* (L_3, L_4 and adults, sum), adult *A. crassus*, and encapsulated *A. crassus* in all sampled eels. Parameters are given for all eels (overall) and separately for each sampling month and each encapsulation status.

		Sum			Larvae			Adults			Capsules
	n	P	I	A	P	I	A	P	I	A	P
Overall	38	89	8.7 ± 1.6	7.8 ± 1.5	79	6.8 ± 1.3	5.4 ± 1.2	55	4.3 ± 1.0	2.4 ± 0.8	39
August	13	92	7.8 ± 2.0	7.2 ± 1.9	85	6.1 ± 2.1	5.2 ± 2.0	46	4.3 ± 1.4	2.0 ± 1.0	46
October	25	88	9.2 ± 2.2	8.1 ± 2.1	76	7.3 ± 1.7	5.5 ± 1.5	60	4.3 ± 1.3	2.6 ± 1.0	36
NC†	19	100	10.2 ± 2.6	10.2 ± 2.6	84	7.2 ± 2.0	6.1 ± 1.8	79	5.1 ± 1.6	4.1 ± 1.4	0
C†	13	100	5.6 ± 1.5	5.6 ± 1.5	92	5.2 ± 1.6	4.8 ± 1.5	38	2.0 ± 0.5	0.8 ± 0.3	100

n, number of eels examined, NC, non-encapsulating eels, C, encapsulating eels

†includes only individuals for which the encapsulation status could be determined unambiguously

Parasite community

The abundance of *A. crassus* in the swim bladder ranged from 0-46, with 0-35 larvae and 0-25

adult worms (Table III.2). Encapsulated *A. crassus* were found in 39% of the eels. Infection intensity

was higher in NC eels than in C eels and it did not differ between sampling months (Table III.3). The

higher intensity in NC eels was largely driven by the large number of parasites found in NC eels in

October (Tukey's HSD, $z_{NCOct-NCAug}$ = 2.56, $p_{NCOct-NCAug}$ = 0.049, $z_{NCOct-CAug}$ = 3.20, $p_{NCOct-CAug}$ = 0.007, $z_{NCOct-COct}$ = 4.36, $p_{NCOct-COct}$ < 0.001). The abundance of larval *A. crassus* did not differ with encapsulation

status but was higher in October than in August with a significant interaction between sampling month

and encapsulation status (Fig. III.1a, Table III.3). *Post hoc* tests revealed that larval load was higher in

NC eels in October than in NC eels in August (z = 3.56, p = 0.004) and that in October NC eels harboured

more larvae than C eels (z = 2.56, p = 0.049). The abundance of adult *A. crassus* was significantly lower

in C eels than in NC eels and it did not differ between sampling months (Fig. III.1b, Table III.3). Heavier

and larger individuals did not harbour more *A. crassus* or more larval stages for any encapsulation

status or sampling month. The number of adult parasites was positively correlated with eel weight and

TL for the NC group (Spearman's rank correlation test, rho = 0.50, p = 0.028 for both weight and TL)

but not the C group (weight: rho = -0.14, p = 0.64, TL: rho = -0.20, p = 0.50). There was no correlation

between the number of adult parasites and weight or TL in August or October.

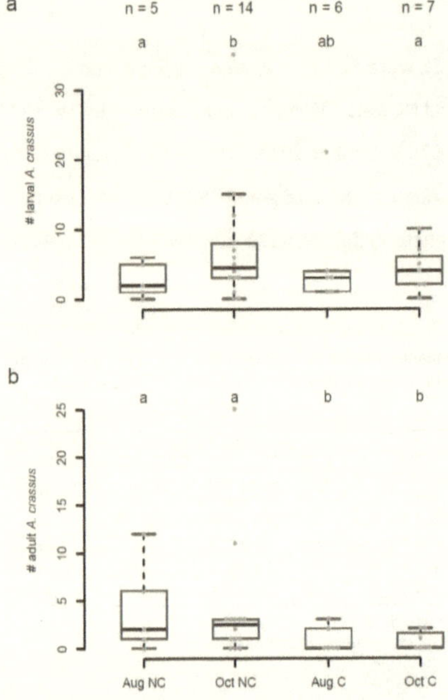

Figure III.1 Infection intensities of (a) all living stages of *A. crassus* and (b) adult *A. crassus* in the swim bladder. Aug and Oct indicate the sampling months, C and NC indicate presence and absence of capsules. Statistically significant values (Tukey's HSD) are indicated with different lower case letters above error bars. Boxes indicate the interquartile range and whiskers extend to 1.5x the interquartile range.

Overall prevalence of the native parasites across both sampling months and encapsulation status ranged from 6% for *Camallanus lacustris* to 78% for *Myxidium giardi* (Table III.4). The invasive *Pseudodactylogyrus* species (*P. bini* and *P. anguillae*) had a prevalence of 100%. Overall mean infection intensities ± SE ranged from 1 for *Diplostomum* sp. to 1.5 ± 0.3 for cestodes (*Bothriocephalus claviceps* and *Proteocephalus macrocephalus*) and *Ergasilus gibbus* and were thus low compared to *A. crassus*.

Table III.3 Statistics for GLM models for a) infection intensity with *A. crassus* and b) gene expression. Only parameters of interest are shown.

Response variable	Group		Month		Group x Month	
	Deviance	p-value	Deviance	p-value	Deviance	p-value
a						
Sum	**20.12**	**<0.001**	2.43	0.12	**5.79**	**0.016**
Larvae	2.22	0.14	**4.33**	**0.037**	**10.88**	**0.001**
Adults	**36.23**	**<0.001**	0.07	0.79	0.02	0.88
b						
Cox1 spleen	0.29	0.17	0.40	0.11	0.01	0.74
Cox1 head kidney	2.29	0.14	**6.56**	**0.012**	1.29	0.27
Epor spleen	0.49	0.56	**7.17**	**0.027**	<0.01	0.99
Epor head kidney	0.38	0.37	0.04	0.76	<0.01	0.97
Mhc IIA spleen	0.95	0.11	**5.49**	**<0.001**	1.19	0.07
Mhc IIA head kidney	0.41	0.44	0.20	0.59	**5.84**	**0.003**
Mhc IIB spleen	0.29	0.44	**2.94**	**0.014**	0.06	0.72
Mhc IIB head kidney	1.09	0.12	0.09	0.65	**2.18**	**0.028**

Group = non-encapsulating vs encapsulating and Month = August vs October

Table III.4 Prevalence (P, %) of parasites other than *A. crassus* in eels that were included in the gene expression analyses. Parameters are given for all eels and separately for each sampling month and each encapsulation status.

	Location	Overall	August	October	NC	C
n		32	11	21	19	13
Cestodes†	intestine	34	36	33	21	54
Camallanus lacustris	intestine	6	9	5	5	8
Pseudodactylogyrus spp.	gills	100	100	100	100	100
Myxidium giardi	gills	78	64	86	79	77
Ergasilus gibbus	gills	34	45	29	21	54
Myxobolus portucalensis	fins	31	73	10	21	46
Diplostomum sp.	eyes	9	18	5	11	8

n, number of eels examined, NC, non-encapsulating eels, C, encapsulating eels
†*Bothriocephalus claviceps* and *Proteocephalus microcephalus*

Infections with *Pseudodactylogyrus* spp., *M. giardi*, and *Myxobolus portucalensis* were categorized into unequally sized intervals, therefore intensities could not be quantified. *Diplostomum* sp. and *C. lacustris* had low prevalences (< 10%) and were excluded from analyses of parasite community composition. The prevalence of cestodes and *E. gibbus* was approx. 2.6 times higher in C group eels compared to NC group eels and that of *M. portucalensis* was approx. 2.2 times higher, though none was significant (Dev = 3.68, p = 0.055 for cestodes and *E. gibbus* and Dev = 2.25, p = 0.13 for *M. portucalensis*; Table III.4). The prevalence of *M. giardi* did not differ between groups. *M. portucalensis* was approx. 7.6 times more prevalent in August than in October (Dev = 12.37, p = 0.0004; Table III.4). The abundance of cestodes and *E. gibbus* was significantly higher in C group eels than NC group eels (Dev = 4.00, p = 0.046 for both). An analysis of similarity revealed moderate differences in parasite community composition between the two months (R = 0.30, p = 0.001, Fig. III.2a), but only minor differences between C and NC group eels (R = 0.09, p = 0.066, Fig. III.2b). *M. portucalensis* and

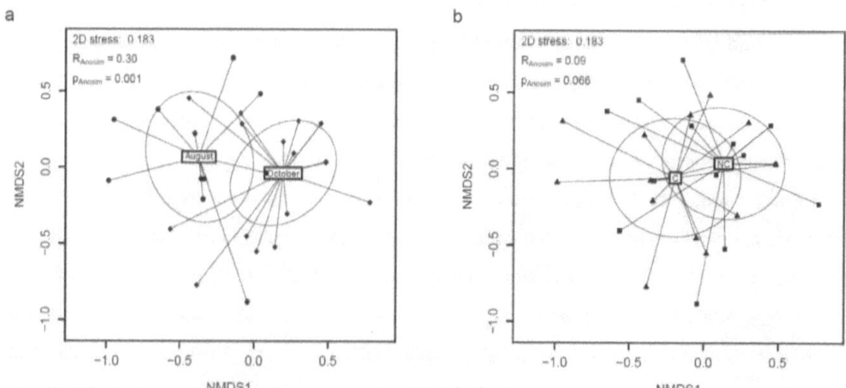

Figure III.2 Non-metric multidimensional scaling plot of parasite communities a) in August (●) and October (♦) and b) for non-encapsulating (NC, ■) and encapsulating (C, ▲) eels.

Pseudodactylogyrus spp. contributed significantly to the parasite community differences between August and October (p = 0.001 for both taxa).

Gene expression

Encapsulation status had no significant effect on relative expression of any studied gene in spleen or head kidney (Fig. III.3, Table III.3). *Cox1* was about 1.8 times more highly expressed in the head kidney of eels sampled in August than in those sampled in October (Fig. III.3b, Table III.3). *Post hoc* tests indicated higher expression in the NC group in August than in October (Tukey HSD, z = 2.64, p = 0.041). Relative expression of *epor* and both *mhc II* genes in the spleen differed between sampling months (Fig. III.3c, e, g, Table III.3). Expression of *epor* was elevated approx. 2.5 fold, *mhc IIA* approx. 2.2 fold, and *mhc IIB* approx. 1.8 fold in August compared to October. For both *mhc II* genes, temporal expression patterns in the head kidney differed with encapsulation status (Fig. III.3f, h).

Expression of both *mhc II* genes in the head kidney correlated negatively with the number of capsules (*mhc IIA*: Dev = 101.8, p = 0.016, *mhc IIB*: Dev = 3.22, p = 0.008). There was no correlation between gene expression and either infection intensity or the number of adult *A. crassus* in either organ. For *mhc IIA* and *mhc IIB*, there was a strong overall correlation between relative expression levels (Spearman's rank correlation, ρ = 0.749, p < 0.001). The correlation remained highly significant when analysing the spleen and the head kidney separately (spleen: rho = 0. 529, p < 0.001, head kidney: rho = 0.542, p < 0.001).

Discussion

Natural populations of European eels are infected by a wide variety of macroparasites (Jakob et al. 2016). The number of parasite species we observed in eels of Lake Müggelsee was comparable to those reported from other European locations (e.g. Sures & Streit, 2001, Jakob et al., 2009, Gérard et al., 2013). Similar to those reports, the invasive parasites *Anguillicola crassus* and *Pseudodactylogyrus* spp. were the most prevalent. Prevalence and mean infection intensity of *A. crassus* were in the upper range of those reported across Europe (e.g. Audenaert et al., 2003, Jakob et al., 2009, Gérard et al., 2013, Becerra-Jurado et al., 2014). We found an increase of larval abundance from August to October. Seasonal dynamics have been reported for several locations (e.g. Lefebvre et al., 2002, Schabuss et al., 2005, Norton et al., 2005) although earlier studies failed to detect such a pattern (Kennedy & Fitch, 1990, Möller et al., 1991, Würtz et al., 1998). Infection intensity and abundance of *A. crassus* were reported to correlate both positively (Schabuss et al., 2005, Neto et al., 2010, Becerra-Jurado et al., 2014) and negatively with eel size (Fazio et al., 2008b, Barry et al., 2017). Similar to our results, another study reported no such relationship (Norton et al., 2005).

Invasive parasites can impose strong selective pressures on their novel hosts which is expected to lead to adaptation of the host (Penczykowski et al., 2011). We found that European eels encapsulating *A. crassus* had fewer parasites in their swim bladders than eels that didn't encapsulate. Encapsulation is observed in naturally infected Japanese eels, *A. crassus'* native host (Münderle et al., 2006, Heitlinger et al., 2009). The lower infection intensity associated with this response in the European eel suggests that the novel host is adapting to its parasite. The reduction was evident for the abundance of adult *A. crassus* but absent for the abundance of larval stages. Thus, encapsulation may prevent the parasite's development to adulthood rather than its establishment and growth. Adult stages are proposed to have the strongest impact on the European eel (Würtz et al., 1996) and reducing their abundance may diminish adverse effects of severe *A. crassus* infections.

The co-infecting parasite community can determine the outcome of infections

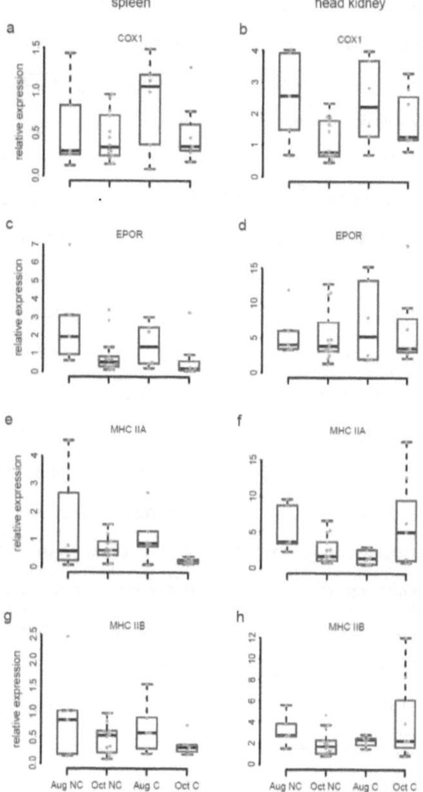

Figure III.3 Expression of target genes relative to the expression of the reference gene β-actin in the spleen (a, c, e, g) and the head kidney (b, d, f, h). Aug and Oct indicate the sampling months, C and NC indicate presence and absence of capsules.

and its impact on the host (Benesh & Kalbe, 2016, Johnson & Hoverman, 2012, Abbate et al., 2018). Co-infecting parasite species can interact with each other either directly via competition, e.g. for resources, or indirectly, e.g. via host immune response, which can suppress or facilitate co-infections (Poulin, 1999, Pedersen & Fenton, 2007). We found abundances of two native parasites to be higher in eels encapsulating *A. crassus*, suggesting that defending against *A. crassus* may ease the establishment of native parasites.

The number of capsules was negatively correlated with *mhc II* expression among eels encapsulating *A. crassus*. Western blot analysis of the antibody response of experimentally infected

European eels suggests that antigens of adult parasites trigger an adaptive immune response (Knopf et al., 2000). More frequent encapsulation may reduce such a response by reducing the number of adult worms. Hence, increased *mhc II* expression may indicate higher susceptibility. Similarly, *mhc IIB* expression in three-spined sticklebacks was positively correlated with parasite load (Wegner et al., 2006). Here, *mhc II* expression in the European eel did not correlate with adult *A. crassus* load, however it may in fact be correlated more strongly with encapsulation and the number of adults than was detected. Our primers do not discriminate the alleles of European eel *mhc II* (Bracamonte et al., 2015). European eels contain at least four expressed *mhc IIA* and *mhc IIB* alleles and if *A. crassus*-specific alleles exist, their change in expression may be masked by expression changes of other alleles induced by the other parasites. The need to respond to more abundant native parasites could also explain why we did not find a difference in *mhc II* expression between encapsulating and non-encapsulating eels.

Adult *A. crassus* are sanguivorous (De Charleroy et al., 1990). Thus a high load of adults may stimulate erythrocyte production in eel hosts. However, expression of *epor*, a gene involved in erythropoiesis, did not differ between encapsulating and non-encapsulating eels. The energy-associated gene *cox1* also did not differ in expression between the groups and expression of both genes was not correlated with the number of capsules nor the number of adult *A. crassus*. The absence of an association between the expression of these genes and the presence or number of capsules may indicate that encapsulating *A. crassus* does not provide energetic benefits over harbouring more adults during the continental phase of the European eel. Alternatively, the higher abundance of native parasites (see above) may countervail any energetic benefits of harbouring fewer adult *A. crassus*.

Our results are in contrast to those reported from experimental infections under controlled conditions. Using a candidate gene approach, Fazio et al. (2009) found increased haemoglobin α expression with increasing parasite biomass. Whole-transcriptome gene expression analyses indicated reduced cell respiration (Bracamonte et al., in press) and the induction of both innate and adaptive immune responses in the presence of larvae (Bracamonte et al., 2019, Bracamonte et al., in press). Increased immune gene expression in infected eels was also observed after acclimation to a common, stress-free environment (Schneebauer et al., 2017). The differences between those and our results may be due to additional biotic and abiotic factors, such as the presence of other parasites or seasonal water temperature changes, that can influence the response to *A. crassus* in the wild. However, all of our individuals were infected with *A. crassus* and we cannot exclude the possibility that the mere presence of *A. crassus* determines the physiological status and the initiation of an immune response regardless of infection intensity. Extending the analysis to non-infected individuals with healthy and highly damaged swim bladders, which result from severe *A. crassus* infections and reduce eel survival during unfavourable conditions (Lefebvre et al., 2007), may offer further insight into the importance of *A. crassus* on the physiological status of wild continental European eels.

The presence of encapsulation of *A. crassus* may be spatially variable. At a site in Lake Müggelsee that is less than 1 km away from our sampling site, the prevalence of capsules in August was only 8% (almost six times lower, n = 12, pers. observation). The European eel is (nearly) panmictic (Als et al., 2011, Baltazar-Soares et al., 2014), therefore the differences in encapsulation prevalence are unlikely to be a result of genetic differentiation, especially on such a small geographic scale. The two sites differ in environmental factors, e.g. parasite community composition, substrate, and shore line structure, and these environmental differences may contribute to the lower encapsulation prevalence observed at the nearby site.

Independently of *A. crassus* or its encapsulation, expression of all genes varied between August and October indicating that multiple environmental factors affect expression of the selected genes. All genes showed higher expression in August than in October in one of the two tissues. Fish are ectothermic, hence colder water in October could be one factor driving the difference (Logan & Somero, 2010, Brown et al., 2016). Parasite community composition also differed between sampling dates and this could lead to variation in gene expression. Translocated sticklebacks adjust the expression profiles of immune genes to those of the local population, which is likely a response to encountering different parasite communities (Stutz et al., 2015). Temporally changing parasite communities can be expected to cause similar adjustments of the response.

In conclusion, the invasive parasite *A. crassus* was one of the most prevalent parasites of eels of Lake Müggelsee. The presence of encapsulation, a frequent response in the parasite's native host, was associated with reduced infection intensity, particularly of blood-feeding adult *A. crassus*, indicating ongoing adaptation by the novel host and that it may follow the same trajectory as in the native host. The lower number of adult *A. crassus* may reduce the necessity for activating the adaptive immune system as suggested by the negative relationship between *mhc II* gene expression and the number of capsules. However, adapting to the novel parasite may alter existing interactions and reduce the ability to cope with native parasites, which may counteract any advantage provided by encapsulation of *A. crassus*. Together with a potentially weak effect of *A. crassus* on the continental stage of eels, these changes to the parasite community may explain the lack of response in the energy-related and the haematopoietic gene.

GENERAL DISCUSSION

The translocation of organisms, accidentally or not, has introduced non-native parasites into foreign habitats (Daszak et al., 2000, Peeler et al., 2011). Livestock trade, pet trade, and restocking of animals for sporting reasons or conservation are the main drivers for large-scale animal movement (Cunningham, 1996, Fèvre et al., 2006, Kock et al., 2010). Insufficient and inadequate disease control of translocated animals are the main source of parasite introductions into foreign habitats (Fèvre et al., 2006, Kock et al., 2010, Tompkins et al., 2015). Parasites that get introduced into the wild can infect local hosts that may be naïve, either of the same species or of different species through host switching, and establish self-sustaining, invasive populations (Lymbery et al., 2014). Aquatic organisms are mainly translocated for aquaculture and ornamental fish trade (Peeler & Feist, 2011, Adlard et al., 2015). Co-translocated parasites can spread easily from aquaculture facilities to adjacent aquatic environments due to insufficient prevention of parasite escape (Peeler & Feist, 2011, Peeler et al., 2011, Tompkins et al., 2015). As a result, more introduced parasites have been identified in fish than in any other vertebrate family (Tompkins et al., 2015).

Invasive parasites have caused epidemics and disease emergence worldwide and several of these parasites have been devastating to their new host species (Daszak et al., 2000, Peeler et al., 2011, Adlard et al., 2015, Staley & Bonneaud, 2015). Avian malaria (*Plasmodium relictum*) caused a massive decline of Hawaiian low-land birds when it was introduced at the beginning of the 20th century (van Riper et al., 1986). The introduction of the white-nose syndrome-causing fungus *Pseudogymnoascus destructans* has resulted in population collapses of North American bat species (Frick et al., 2010). Chytridiomycosis, a disease caused by the chytrid fungus *Batrachochytrium dendrobatidis*, causes the ongoing worldwide massive decimation of amphibian populations (Scheele et al., 2019). Invasive fish parasites have also been linked to population declines (Peeler et al., 2011). Rainbow trout populations collapsed after the introduction of the myxozoan *Myxobolus cerebralis* into North American rivers (Bergersen & Anderson, 1997). In Europe, the parasite *Sphaerotecum destruens* is associated with high mortality in various fish species (Ercan et al., 2015) and the nematode *Anguillicola crassus* is implicated in the decline of the European eel (Drouineau et al., 2018). The many examples of invasive parasites raise the question why novel hosts are more susceptible than native host. It has been suggested that long-standing co-evolution and adaptation to the parasite in the native host resulted in a more effective immune response and thus the ability to confine an infection. In contrast, novel hosts lack such immunity (Taraschewski, 2006, Mastitsky et al., 2010, Peeler et al., 2011). However, the mechanisms underlying this assumption and the processes involved in higher susceptibility in novel hosts are not well understood.

Angillicola crassus is a swim bladder parasite of eels native to the Japanese eel (*Anguilla japonica*). It was introduced into the European eel (*A. anguilla*) population in the early 1980s (Kirk, 2003) and it is implicated in the severe population decline of the European eel and its current status as critically endangered (Jacoby et al., 2015, Miller et al., 2016, Drouineau et al., 2018). Infection intensity in naturally and experimentally infected Japanese eels is lower than in European eels (Audenaert et al., 2003, Knopf & Mahnke, 2004, Lefebvre & Crivelli, 2004, Münderle et al., 2006, Heitlinger et al., 2009, Weclawski et al., 2013). Infected European eels display higher oxygen consumption (Palstra et al., 2007), altered swimming patterns (Palstra et al., 2007, Newbold et al., 2015), and reduced swim bladder function (Würtz et al., 1996, Barry et al., 2014) compared to uninfected conspecifics. Furthermore, gene expression indicates that osmoregulation and haematopoiesis may be compromised in infected European eels (Fazio et al., 2009), that a large array of processes is responsible for the parasite's effect on swim bladder function (Schneebauer et al., 2017), and that infection may alter the silvering process (Fazio et al., 2012, Pelster et al., 2016). The effects of *A. crassus* on the European eel suggest that the parasite interferes with the spawning migration and reproduction. In contrast, no negative effects of infection have been reported for the Japanese eel.

The first aim of the book was to identify genes and processes involved in the early response to the invasive parasite *A. crassus*, i.e. during migration of parasite larvae and following establishment in the swim bladder but before development into adult stages. The goal was then to determine if differences in these early response between the European eel and the Japanese eel could explain the higher susceptibility observed in the former (Knopf, 2006). For this, both European eels and Japanese eels were infected experimentally with *A. crassus* and differential gene expression in the head kidney was determined between infected and sham-infected control individuals at 3 and 23 dpi using RNA-seq. Additionally, differential gene expression in the spleen was determined for European eels at 3 dpi to identify differences in the response to *A. crassus* infections among immune organs. Experimental infections resulted in similar infection intensities in European eels and Japanese eels at 23 dpi, when migration of *A. crassus* to the swim bladder was completed. European eels altered expression patterns of both immune and non-immune genes. In the spleen, a major secondary immune organ, only a few genes showed differential expression at 3 dpi. Over-represented GO terms were mostly related to an inflammatory response and cytokine production. In contrast, in the head kidney about five times as many genes were differentially expressed at 3 dpi. Over-represented GO terms indicated that they were involved in a much more diverse set of processes, including innate and adaptive immune responses, energy generation, and kidney function. Expression of genes involved in immune responses and metabolism was still altered at 23 dpi in the head kidney. For the Japanese eel, there were only a few differentially expressed genes in the head kidney at both 3 and 23 dpi and, unexpectedly, they did not indicate activation of a systemic immune response. Also, genes involved in metabolic processes and kidney function were not affected with the exception of the erythropoietin receptor (*epor*) involved in red blood cell maturation and proliferation.

The changes in gene expression that were observed in the head kidney and the spleen of experimentally infected versus control European eels in the present study suggest that both organs participate in the response to *A. crassus*. However, the head kidney may have a more prominent role in the response. The head kidney is a complex organ of fish that combines immune and non-immune functions. It serves as primary and secondary immune organ and as adrenal gland (Whyte, 2007). Furthermore, the eel head kidney is more similar to a mid kidney (Tesch, 2003). As such, it may additionally be involved in renal processes (Whyte, 2007) which is supported by the present data.

A larger disruption of immune and non-immune processes was observed in the head kidney of experimentally infected European eels compared to Japanese eels in this study, suggesting that the European eel responds more sensitively to infections with larval *A. crassus*. In the European eel, innate immune processes were activated at both time points and adaptive immune processes were activated at 23 dpi. Such a progression from innate to adaptive immune processes was also observed for mice infected with the parasite *Eimeria falciformis* (Ehret et al., 2017). However, the immune activation in the European eel observed here did not lead to a reduction of parasite load compared to the Japanese eel for which no such immune activation was observed in the present study. Thus, the immune response mounted by the European eel may not be effective to defend against *A. crassus*. Similar immune activation in susceptible novel hosts but not in more resistant hosts was described for frogs and toads infected with the chytrid *B. dendrobatidis* (Ellison et al., 2015, Poorten & Rosenblum, 2016, Eskew et al., 2018) and for *Myotis* bat species infected with the fungus *P. destructans* (Field et al., 2015, Davy et al., 2017).

Despite the activation of immune processes being indicative of higher susceptibility, an effective immune response is indispensable for coping with infections (Murphy & Weaver, 2017). In contrast to the observations here, the more resistant species of amphibians and bats had reduced parasite load compared to the susceptible species, suggesting that immune responses were activated at some point during the infection (Ellison et al., 2015, Field et al., 2015, Poorten & Rosenblum, 2016, Davy et al., 2017, Eskew et al., 2018). The Japanese eel reduces the infection intensity over the course of infection (Weclawski et al., 2013) which indicates that an immune response is mounted later, probably after adult parasite stages appear. A late immune response could explain why there was no clear signal of immune response here both at 3 and 23 dpi when only larval stages were present. It could also account for the similar infection intensities between eel species observed here, despite previous reports of lower intensities in the Japanese eel several weeks post-infection (Knopf & Mahnke, 2004, Knopf &

Lucius, 2008, Weclawski et al., 2013). The timing of and the mechanisms involved in the Japanese eel's immune response are still unknown. Antibodies are produced eight weeks after injection with antigens of adult *A. crassus* (Nielsen, 1999) and nine weeks after experimental infection (Knopf & Lucius, 2008), but the antibody content does not relate to infection intensity, suggesting that they are not an essential part of the response (Knopf & Lucius, 2008). Elucidating the timing and progression of the response by native host species and the processes involved can inform why they are capable of resolving or reducing an infection while novel host species are not.

An immune response, even a successful response, comes at considerable costs to the host itself (Graham et al., 2005). The cytotoxicity inherent to immune responses causes damage not only to the parasite, but also to host tissue, which requires subsequent repair (Graham et al., 2005, Schulenburg et al., 2009). Furthermore, mounting an immune response carries energetic and metabolic costs resulting in increased breakdown of glucose, carbohydrates, proteins, and lipids (Lochmiller & Deerenberg, 2000, Viney et al., 2005). Metabolism and transport of all of these compounds were altered in the head kidney of the experimentally infected European eels in the present study. In contrast, resistant species may optimize their resources by mounting a strictly regulated and hence more efficient immune response only when parasite-inflicted damage outweighs the costs of mounting a response (Viney et al., 2005, Sadd & Schmid-Hempel, 2009). This optimization strategy could account for the lack of an immune response to larval stages of *A. crassus* in the Japanese eel and the absence of an effect on metabolic processes observed here. The difference in response strategies between the European eel and the Japanese eel could thus indicate debilitation of infected European eels. This may result in early disease development in the European eel but not the Japanese eel, despite similar infection intensities. If the resource use during the early stages of infection, i.e. when only larval stages are present, weakens the European eel enough to render it incapable of coping with the presence of adult parasites, the early, and thus far overlooked (Nielsen, 1999, Knopf et al., 2000, Knopf & Lucius, 2008, but see Fazio et al., 2009) response to larval parasites may contribute substantially to the higher susceptibility of the European eel (Knopf, 2006, Taraschewski, 2006). Transcriptome-wide gene expression patterns in amphibians infected with the invasive chytrid fungus *B. dendrobatidis* and in bats infected with the invasive fungus *P. destructans* causing white-nose syndrome indicate that the activation of the immune system in susceptible but not resistant species went along with alterations of metabolic processes only in the susceptible species (Ellison et al., 2015, Field et al., 2015, Poorten & Rosenblum, 2016, Davy et al., 2017, Eskew et al., 2018). Activating a prolonged and ineffective immune response may thus deplete resources and inflict self-harm without alleviating damage and resource use by parasites in novel and susceptible species. Hence, the excessive disruption of metabolism may generally contribute to higher susceptibility in these hosts. A comprehensive view of the impact of invasive parasites on novel hosts may only be achieved by paying more attention to processes other than the immune response.

Interactions between hosts and parasites allow for reciprocal adaptation over time (Anderson & May, 1982, Woolhouse et al., 2002). A major caveat of this and previous comparative studies of *A. crassus* infections in the European eel and the Japanese eel (e.g. Nielsen, 1999, Knopf & Mahnke, 2004, Knopf & Lucius, 2008) is the use of parasites of European origin only for experimental infections aiming to identify why the European eel is more susceptible. *Anguillicola crassus* of European origin have started to differentiate both genotypically (Heitlinger et al., 2014) and phenotypically from those of Asian origin and this was proposed to indicate ongoing adaptation of the parasite to the novel host (Weclawski et al., 2013, Weclawski et al., 2014). The life cycle of *A. crassus* can be completed in less than 2 months under laboratory conditions, i.e. 20 °C water temperature (De Charleroy et al., 1990), and water temperatures above 10 °C are already likely to support completion of the life cycle (Bonneau et al., 1991, Knopf et al., 1998). This suggests that *A. crassus* can have several generations per year also in nature. Even assuming a conservative number of two generations per year would result in approximately 70 generations since its introduction into Europe, as opposed to 3-5 generations for the European eel. Differences in generation times provide vastly different evolutionary rates and control the adaptation potential of the interacting species. According to evolutionary theory, parasites are expected to be adapted to their local hosts on average and maladapted to foreign hosts (i.e. local adaptation; Anderson & May, 1982, Kaltz & Shykoff, 1998). Hence, European *A. crassus* may have become maladapted to the Japanese eel host. This could influence how the Japanese eel responds to *A. crassus* of European origin compared to *A. crassus* of Asian origin. The response pattern of the Japanese eel reported in the present study may represent one that is induced by a parasite not fully capable of exploiting its host. Determining the Japanese eel's response to *A. crassus* of Asian origin may provide a more comprehensive view on the mechanisms underlying the different susceptibilities of the European eel and the Japanese eel.

Similar to parasites adapting to novel hosts, hosts can adapt to novel parasites which may allow for species persistence and recovery (Penczykowski et al., 2011). A textbook example is the adaptation of rabbits in Australia and Europe to the release of the myxoma virus in the early 1950s. Rabbits on both continents increased in resistance to the virus within few generations and the virus changed from an epidemic killing off > 90% of the population to an endemic state with fatalities of 20-30% (Kerr, 2012). A North American rainbow trout population may be acquiring resistance to whirling disease after the introduction of the disease causing myxospore in the 1950s. In experimental infections, progeny of older cohorts suffered larger pathological damage than progeny of younger cohorts (Miller & Vincent, 2008). The Hawaiian low-land amakihi population is recovering from a severe collapse after avian malaria was introduced into the archipelago. Experimental infections suggest improved survival and physiological condition but no difference in infection intensity for low-land amakihi compared to susceptible high-land amakihi, indicating acquired tolerance rather than resistance (Atkinson et al., 2013). The second aim of the book was to determine if the European eel is adapting to *A. crassus*. Encapsulation of *A. crassus*, the response observed in the native Japanese eel host (Münderle et al., 2006, Heitlinger et al., 2009) has been reported for the European eel, although only sporadically (Molnár, 1994, Audenaert et al., 2003). However, the proportion of encapsulated *A. crassus* raised from 0% in 1990 to 20% in 1997 and 2000 in Flanders, Belgium and at the same time the proportion of larval stages was cut in half (Audenaert et al., 2003). However, the authors did not consider encapsulation beneficial but rather suggested that it may cause further damage to the swim bladder, even though encapsulation is considered the reason for lower susceptibility of the Japanese eel (Knopf, 2006). Here, the aim was to examine if encapsulation of *A. crassus* reduces the infection intensity in the European eel and if this potentially adaptive process acts on genes responding to experimental infections. For this, infection intensity, parasite community composition, and expression of candidate genes selected based on the transcriptome-wide expression studies were determined in naturally infected European eels encapsulating *A. crassus* and compared to those of co-occurring European eels not encapsulating *A. crassus*. Infection intensity, particularly that of adult parasites was reduced in individuals encapsulating *A. crassus*. However, they suffered from higher abundance of two native parasites. Expression of selected immune genes that were affected by experimental infections (*mhc IIA* and *IIB*) was negatively correlated with the number of encapsulated *A. crassus* in naturally infected European eels, providing evidence for an altered immune response. Surprisingly, the reduced number of *A. crassus* did not result in altered expression of an energy-related gene (*cox1*) or *epor*, one of the few differentially expressed genes in experimentally infected Japanese eels. The pattern of markedly reduced adult *A. crassus* with encapsulation was present in summer and autumn but both gene expression and parasite community composition differed between seasons.

Increased encapsulation frequency (Audenaert et al., 2003) and reduced parasite load, particularly of adult stages, observed here suggest that encapsulation of *A. crassus* is an adaptive process. Adult *A. crassus* appear to be the most costly parasitic stage (Würtz et al., 1996) and reducing their load may improve host performance and ultimately support population recovery of the European eel. However, the present study assessed the relationship of encapsulation and parasite abundance at a single site within the continent-wide distribution range of the European eel (Tesch, 2003). Further observations across the entire distribution range are needed to consolidate the hypothesis that encapsulation is indeed adaptive. Observing a similar pattern in the Japanese eel could also support the adaptive value of encapsulating *A. crassus*. Only up to 22% of contain encapsulated *A. crassus* (Münderle et al., 2006, Heitlinger et al., 2009), opposed to 39% of European eels here, and it is not known if responding Japanese eels actually harbour fewer parasites than non-responders. This raises the question if different infection intensities between the two eel species are really the result of clearing and encapsulating *A. crassus* and if there are additional factors contributing to lower infection intensity and prevalence in the Japanese eel. These could include behavioural differences, for example avoiding feeding on infected prey, or a lower abundance of infective stages in Asian waters compared to European waters. *Anguillicola crassus* has expanded its intermediate and paratenic host range following invasion into Europe (see below) and this could promote higher abundance. The introduction of *A. crassus* into the American eel population approximately 10 years after its introduction into Europe (Barse & Secor, 1999, Hein et al., 2014) provides an opportunity to determine if coping with *A. crassus* always involves the same mechanisms. Encapsulation in the American eel and its association with lower infection intensities would support this hypothesis.

Encapsulation and lower infection intensity suggest that the European eel has evolved some degree of resistance to *A. crassus*. Increased resistance is typically the result of mounting an immune response that is specific to the infecting parasite (Raberg et al., 2007, Schneider & Ayres, 2008). In a frog population exposed to an invasive chytrid parasite, increased resistance was associated with a temporally more restricted, more precise expression of immune genes and lesser modifications of metabolic genes compared to populations that have not evolved resistance to the parasite (Grogan et al., 2018). Ecotypes of three-spined sticklebacks that are more or less resistant to particular parasites also differ in their immune and metabolic gene expression profiles with the more resistant ecotype upregulating immune genes and the less resistant ecotype upregulating metabolic genes (Lenz et al., 2013). A rainbow trout strain resistant to *Myxobolus cerebralis* increases expression of genes belonging to a different set of immune pathways than a susceptible strain (Baerwald, 2013). Gene expression of responding vs non-responding European eels in this study suggested only a weak association with an adaptive immune response and no difference in metabolism. However, encapsulation may be driven by the innate immune system (Dezfuli et al., 2015) and genes such as those associated with macrophage stimulation or the complement component, which were involved in the response in the spleen (see chapter I), the head kidney (see chapter II), and the swim bladder (Schneebauer et al., 2017), may have altered expression in responding eels. In the three-spined stickleback, *mhc IIB* expression is indicative of susceptibility rather than resistance as expression was strongest in less resistant and more heavily infected genotypes (Wegner et al., 2006, Wegner et al., 2007) and the negative correlation of *mhc II* expression with the number of capsules here suggests that this may also be the case for the European eel. Besides expression patterns, selection for resistance may have altered *mhc II* genetic composition. *Mhc* alleles or haplotypes, for example, are associated with increased and reduced resistance to specific parasites in various species (e.g. Bonneaud et al., 2006, Dionne et al., 2009, Eizaguirre et al., 2011, Loiseau et al., 2011, Kaufmann et al., 2017). Resistance *mhc II* haplotypes increased in frequency within two generations of three-spined sticklebacks experimentally infected with one of two parasites (Eizaguirre et al., 2012b). It is possible that resistance alleles to larval *A. crassus* are being selected and spreading across the European eel population. In that case, only those individuals carrying resistance alleles may be able to respond to and encapsulate the parasite larvae.

The impact of *A. crassus* on the European eel appears to vary under different environmental conditions and over the European eel's life cycle. *Anguillicola crassus* may have only a minor effect on the continental yellow eel stage during normal conditions (Höglund et al., 1992, Kelly et al., 2000, Lefebvre et al., 2013). Encapsulation may thus only provide a small and undetectable advantage during this stage which may explain why responding and non-responding European eels did not differ in metabolic gene expression in the present study. During experimentally induced hypoxia, however, naturally infected eels with the highest parasite load and those with most highly damaged swim bladders died first (Molnár, 1993, Lefebvre et al., 2007). Thus, being capable of responding efficiently to the infection and reducing parasite load by encapsulating *A. crassus* instead of mounting an ineffective immune response, as suggested by experimental infections (this study, Knopf et al., 2000), may be favourable under stressful conditions. This could then improve the performance of individuals encapsulating *A. crassus* which may lead to detectable changes of metabolic processes and expression of associated genes under such conditions. Alternatively, the ecological context may differ for responding and non-responding eels in the complex natural environment which may counteract any benefits of increased resistance (see below). For silver European eels, encapsulation could provide a considerable advantage. Silver eels do not feed and their energy stores need to support the high metabolic costs of the spawning migration, maturation, and reproduction (van den Thillart et al., 2004, Palstra & van den Thillart, 2010). *Anguillicola crassus* infections, however, increase the costs of migration and may not leave enough energy to complete reproduction (Palstra et al., 2007). Therefore, it has been hypothesized that the majority of spawners contributing to the following generation are those that spent their entire life in brackish or marine waters where infection probability is considerably lower than in fresh water (Jakob et al., 2009, Dezfuli et al., 2014, Wysujack et al., 2014). Lower infection intensities in responders reported here may reduce the additional migrating costs of these eels and ensure successful reproduction also of freshwater-inhabiting eels. Identifying the differing costs of harbouring an infection, encapsulating *A. crassus*, and coping with severe parasite-induced damage in nature for both the sedentary yellow eel and the migrating silver eel may give further insight into the advantages of the different response phenotypes and whether encapsulation may promote population recovery.

Novel parasites can alter interactions within the species community and this may determine the impact of the parasite on the host (Johnson & Paull, 2011, Goedknegt et al., 2016). Co-infections are common in nature (e.g. Knudsen et al., 2003, Eizaguirre et al., 2011) and the European eel is also simultaneously infected by multiple parasites throughout its distribution range (Kennedy, 2001, Sures & Streit, 2001, Fazio et al., 2008b, Jakob et al., 2009, Gérard et al., 2013, Dezfuli et al., 2014, this study). The outcome of co-infections commonly differs from single infections (Graham, 2008, Johnson & Hoverman, 2012, Benesh & Kalbe, 2016, Betts et al., 2016, Karvonen et al., 2019). Different infection outcomes can arise due to direct interactions among the parasites or indirect interactions via host response, the latter being related to the activation of an immune response (Poulin, 1999, Graham, 2008, Cézilly et al., 2014, Hafer & Milinski, 2016). An inadequately activated immune response, as observed in novel hosts including the European eel (see above), can be expected to facilitate infection by multiple parasite species. Consequently, the introduction of an invasive parasite may disrupt the response to the native parasite community. Both rainbow trout and blue mussel experimentally infected with a non-native parasite had reduced survival when simultaneously infected with a native bacterium (Densmore et al., 2004, Demann & Wegner, 2019). In addition, co-infection tended to reduce phagocytic activity in blue mussels, a measure of immunocompetence (Demann & Wegner, 2019). Here, European eels responding to *A. crassus*, rather than non-responders, suffered from higher abundance of some native parasite species. This suggests that not only co-infection with, but also adapting to an invasive parasite may change interactions between the novel host and its native parasite community, although it remains unknown how *A. crassus* infection itself affects European eel-native parasite interactions, irrespective of the response. Nonetheless, the higher abundance of native parasites in responders could contribute to the lack of observed benefits that would be expected in case of adaptation (see above). However, the association observed here does not indicate causality. Encapsulation may increase the abundance of native parasites but the higher abundance may also facilitate encapsulation, e.g. via cross-reactive immune responses. It is further possible that responding and non-responding European eels predominantly occupy different microhabitats or have different feeding strategies which may expose them to a differing parasite community and promote or hamper encapsulation of *A. crassus*.

The environment may influence the European eel's capacity to respond to *A. crassus*. Despite having suggested that encapsulation may be an adaptive response to *A. crassus* (see above), the present study indicated that the prevalence of capsules was only 8% (n_{eels} = 12) at a site within Lake Müggelsee less than 1 km apart from the main sampling site which had an encapsulation prevalence of 46% (n_{eels} = 13) for the same month. Observing such differences raises the question under which conditions encapsulation occurs. Does the environment define the ability to encapsulate or do all combinations of environmental factors allow for encapsulation but only for some it is advantageous? The two sites differed in the structure of the substrate and the shoreline and eels harboured different parasite communities. The parasitic copepod *Ergasilus gibbus*, for example, was almost ten times more abundant on the gills of European eels from the site 1 km away than of those from the main sampling site of the study. It is likely that also the free-living species community and abiotic factors differed between sites. Although these differences could affect both the ability to encapsulate and its benefits, they hint at the importance of environmental conditions for observing one or the other response phenotype. Heterogeneous environments provide a patchy network of selective pressures and offer the possibility for populations to adapt locally. Three-spined stickleback populations have adapted to their habitat-specific parasite faunas which is reflected in differing immune responses to and infection intensities with particular parasite species (Kalbe & Kurtz, 2006, Eizaguirre et al., 2012a, Kalbe et al., 2016). The European eel encounters a large diversity of environments across its distribution range and local single-generation selection to abiotic factors has been proposed despite overall panmixia (Pujolar et al., 2014). Different environments could lead to locally variable single-generation selection on the optimal response to *A. crassus*. Unlike in species with discrete (meta-)populations, this would slow down adaptation. For encapsulation to provide a means of adaptation to *A. crassus*, it would need to be the optimal response on a population-wide scale. The interaction of *A. crassus* with its intermediate and paratenic hosts may potentially also influence the response of the European eel to *A. crassus*. Unfortunately, to date nothing is known about the intermediate and paratenic hosts' response and its effect on the eel's response and this merits future consideration. For *Schistosoma mansoni*, a parasite with a complex life cycle similar to that of *A. crassus*, the infection success in the final mouse host depended strongly on the genetic background of the intermediate snail host (Zavodna et al., 2008). Similar differences in infectivity to the eel may occur for *A. crassus* transmitted from different intermediate and paratenic host species and populations and this may contribute to the differences of encapsulation among sampling sites that were observed here. If fending off an immune response by a previous host weakens *A. crassus* it may facilitate encapsulation by the eel. Different defence capabilities of intermediate and paratenic host species could also lead to different parasite abundances among them. Feeding preferentially on resistant hosts could then result in higher encapsulation and lower infection intensity compared to feeding on tolerant or susceptible hosts. The proportion of fish in the diet of European eels influenced infection intensity with *A. crassus*, however one study found higher parasite load in predominantly fish eating eels (Pegg et al., 2015) while another found lower parasite load when fish were the dominant food item (Barry et al., 2017). Elucidating how processes in the various hosts interact to shape infection outcomes may potentially advance our understanding of infection dynamics and adaptive potential of novel hosts as well as novel parasites.

Despite the many examples of introduced parasites that were detrimental to novel host species, novelty does not necessarily mean higher susceptibility. There is likely a considerable proportion of translocated parasite species that does not manage to acquire novel hosts in the introduced region (Pizzatto et al., 2012, Sheath et al., 2015, Rauque et al., 2018). Though determining the reasons that prevent host switching is difficult and may be the sum of many processes, resistance of potential novel hosts is likely to contribute substantially to the lack of parasite establishment. Similarly, invasive free-living species are exposed to a plethora of potential novel parasites that are native to the invaded region. However, host switching of these parasites to the invading species does not always occur (Telfer et al., 2005, Kopp & Jokela, 2007, Pizzatto et al., 2012, Nelson et al., 2015, Mori et al., 2019), again suggesting that novelty does not necessarily translate into higher susceptibility. Just as novel hosts can be susceptible or resistant, native hosts too can be more susceptible or more resistant. Endemic parasites are found everywhere. They can cause epidemics in their native hosts and their prevalence and infection intensity differs among host populations indicating that there are at least episodes of high susceptibility (Dobson & Foufopoulos, 2001, Duffy & Sivars-Becker, 2007, Gibson et al., 2016, Weber et al., 2017). These observations raise two important questions: 1) why are some hosts susceptible to novel parasites while others are not and 2) why are some but not all parasites capable of switching hosts and establish? These questions are not independent of each other as infection and virulence are traits jointly controlled by hosts and parasites (Woolhouse et al., 2002). *Anguillicola novaezelandiae*, a parasite infecting short-finned eels (Lefebvre et al., 2004), has been introduced into a lake in Italy where it successfully infected resident European eels. However, it hasn't spread to other locations and shortly after the introduction of *A. crassus* into the same lake *A. novaezelandiae* disappeared (Moravec et al., 1994b). Differences in life cycle have been suggested for the successful invasion by *A. crassus* but not *A. novaezelandiae* (Dangel et al., 2013). Experimental infections with both species suggest that *A. crassus* develops faster in the intermediate host than *A. novaezelandiae* (Bonneau et al., 1991, Moravec et al., 1994a, Dangel et al., 2015). In the final European eel host, maturation time appears to be similar for both parasite species (Weclawski et al., 2013, Dangel et al., 2013, Keppel et al., 2014, Dangel et al., 2015). However, *A. novaezelandiae* develops in synchrony while *A. crassus*, the successful invader, develops asynchronously, thus potentially supplying reproductively active individuals continuously over a longer time period. This could lead to a higher reproductive output and facilitate the spread to other locations (Dangel et al., 2013, Keppel et al., 2014, Dangel et al., 2015). However, *Anguillicola* species have a complex life cycle and they depend on intermediate hosts. Hence, the greater invasion potential of *A. crassus* could also occur at intermediate host level rather than final host level. Furthermore, *A. crassus* employs paratenic hosts in Europe (Thomas & Ollevier, 1992, Kirk, 2003) and in North America (Li et al., 2015, El-Shehabi et al., 2018) which may enhance its invasion potential. Whether *A. novaezelandiae* infects paratenic hosts is not known.

The success of a parasite in invading a novel host may depend on the phylogenetic relatedness of novel and native parasites and hosts. Experimental infections with *A. novaezelandiae* resulted in substantially higher infection intensities in the European eel compared to the Japanese eel, suggesting that the immune response of the Japanese eel mounted against the native *A. crassus* may respond efficiently to the unfamiliar, but related parasite (i.e. immunological cross-reactivity; Keppel et al., 2016). In contrast, for the European eel no parasites closely related to *A. crassus* and *A. novaezelandiae* have been reported (Jakob et al., 2016). Hence, the presence of phylogenetically similar parasites in the native parasite community may hamper the establishment of an introduced parasite. This may, however, not be generalizable. Malaria parasite strains have invaded host species where local strains were present (Schoener et al., 2014, Marzal et al., 2015). Phylogenetic relatedness of hosts, such as the Japanese eel and the European eel, may facilitate host switching of novel parasites. However, *A. crassus* was reported to only infect cyclopoid copepods as intermediate hosts in Southeast Asia, its original range (Nagasawa et al., 1994), but it can infect a suite of intermediate hosts in Europe other than cyclopoid copepods (Kennedy & Fitch, 1990, Kirk, 2003). Also, it uses a wide range of fishes belonging to different families as paratenic hosts in Europe (Haenen & van Banning, 1990, Thomas & Ollevier, 1992, Moravec & Konecny, 1994, Székely, 1994) which has not been described in its native range (Nagasawa et al., 1994).

Limitations of RNA-seq in ecological studies

In the present study, RNA-seq was used to identify genes that may be relevant to the host response to *A. crassus* and those genes were used for assessing the impact of *A. crassus* in naturally infected European eels. However, there are several challenges when inferring ecologically relevant genes and processes from RNA-seq data. Transcription is only the first step in the production of proteins, which ultimately determine what functional processes are involved in a particular response. Post-transcriptional mRNA stability, translation efficacy, and post-translational modifications that affect protein stability and activity can disrupt the correlation of mRNA and protein content (Plotkin, 2010, Evans, 2015). Proteins may have an important function in the response but achieve that by modifying stability and activity, not transcription. This goes undetected when focussing on changes in mRNA content. Similarly, protein content may be unchanged despite changes in mRNA content if mRNA stability is the determining factor. Hence, the changes in mRNA content observed here may not translate exactly into functional changes at the organism level. However, they provide an approximation of genes and processes that may be relevant for the responses to *A. crassus* infections as it is not likely that the content of mRNA that is unrelated to the response is being modified. Supplementing the transcriptional changes with proteomic changes may, however, improve our understanding of the exact processes associated with the response and the differences between the two eel species.

The most highly differentially expressed genes may not be the most important to cope with a certain condition (Evans, 2015). Rather small expression changes in many interconnected genes may provoke the necessary response. However, transcriptome-wide expression studies, such as the present study, lack the power to detect such small changes that are dismissed as background noise. As a result, the differentially expressed genes reported here likely represent a conservative set. Network analyses considering expression patterns among genes known to interact may capture those subtle changes, because they do not rely on differential expression of single genes (Evans, 2015). However, building the networks relies on *a priori* knowledge of gene interactions and they cannot incorporate unknown associations among genes and non-annotated genes, which are common for non-model organisms including eels. In the present study, the majority of genes did not retrieve and annotation and many interconnecting genes may not have been identified. Also, such networks can be highly complex, share core regulatory genes, and lack functional information, making their interpretation difficult (Nikinmaa & Rytkönen, 2012).

Aside from identifying genes that are relevant for the response, identifying their function poses and additional challenge. Most ecologically interesting species are non-model organisms with no or only poorly assembled and functionally annotated genomes (Pavey et al., 2012). As a result, a considerable proportion of genes are of unknown function. Draft genomes of the European eel and the Japanese eel are available, however, they are not or only partially annotated (Henkel et al., 2012b, Henkel et al., 2012a, Jansen et al., 2017) and the most recent assembly is not recommend to be used for large-scale analyses (Henkel, 2019). Cross-species annotation may fail because the genes are too divergent from those of better annotated species or because the genes are not present in the latter (Primmer et al., 2013, Evans, 2015). This may be particularly prominent for non-mammalian organisms that are not closely related to model organisms. These genes of unknown function may, however, be of particular importance for coping with environmental conditions encountered by the organism and for adapting to new ones (Pavey et al., 2012, Evans, 2015). Here, only approximately 50% of differentially expressed genes could be annotated suggesting that unknown genes may potentially play a role in coping with *A. crassus* infections. The vertebrate immune system appears to be well conserved and teleosts share major features with the mammalian immune system (Alvarez-Pellitero, 2008, Uribe et al., 2011, Buchmann, 2012). However, what functions other than the immune response are involved in responding to infections is still poorly understood (Schmid-Hempel, 2003, Viney et al., 2005, Schulenburg et al., 2009) and the non-annotated genes in the present study may be involved in this non-immune, but potentially important response. The inability to identify the most important genes, either due to missing power for detection or missing annotation, could have led to selecting inadequate candidate genes for the study on naturally infected European eels and it could have contributed to the lack of differential expression between responding and non-responding naturally infected eels that was observed here. Inferring co-expression patterns among genes and identifying enriched functions within co-expressed gene clusters could indicate possible functions of non-annotated genes (Stanford & Rogers, 2018). These analyses may also allow to identify genes with only small expression changes which could substantially improve the functional assignment of the co-expressed gene clusters. However, co-expression does not imply that genes belong to the same process or response mechanism. Nonetheless, they provide a good starting point for further functional characterization and association studies. Hence, future studies aiming to identify the responses to *A. crassus* in the European eel and the Japanese eel may profit from incorporating such co-expression analyses.

Cross-species annotation using better annotated genomes has its caveats too. Many of these annotations have been determined in mammals, mostly humans and rodents. They were then used to annotate genomes of other animals which are further used for cross-species annotation. Errors occurring during annotation of one genome can then be carried over to other genomes (Primmer et al., 2013). Although most annotations of the differentially expressed genes here were consistent between RefSeq and the better curated Swiss-Prot, there were some discrepancies which may indicate that incorrect annotations were carried over. Using correctly annotated genomes for cross-species annotation can, however, also introduce errors if the species of interest is only distantly related to the species used for inferring annotation or if they inhabit habitats with different properties that require different functional adaptations (Nikinmaa & Rytkönen, 2012). Teleost fish, including eels, have gone through an additional round of whole genome duplication compared to mammals providing ample material for subfunctionalization, i.e. splitting various functions of a gene on different variants, and neofunctionalization, i.e. evolving new functions different from the original one (Glasauer & Neuhauss, 2014). Human and rodent genes and genomes are far better studied and annotated than any fish species (Primmer et al., 2013) and in the present study, functional annotation was predominantly assigned based on sequence similarity to mammalian genes. This may have caused incomplete functional annotation for example of genes that are related to an aquatic life style and poikilothermy.

Since infections with *A. crassus* affected metabolic and homeostatic processes in the European eel in this study, such incomplete or erroneous assignments may have failed to identify some of the processes involved in the response.

Another difficulty is that the identification of gene functions depends heavily on the context in which it has been studied. Gene functions in one organ may differ from functions in another organ (Primmer et al., 2013). Similarly, functions may differ among developmental stages or in health and disease. The function of many genes during a state of disease was identified for bacterial infections and in cancer cells. However, the processes involved in a response to nematode parasites such as *A. crassus* differ from those to bacteria or cancer (Díaz & Allen, 2007, Murphy & Weaver, 2017) and a gene's function may differ accordingly. As a result, functional annotations for genes that are highly interconnected and involved in various processes are likely to be incomplete which again may lead to processes potentially involved in a response not being identified and the results presented here being conservative.

The European eel has been found to have a higher parasite load of the invasive parasite *Anguillicola crassus* than the Japanese eel, the parasite's native host, and *A. crassus* was reported to negatively affect the health of its novel host. This book found that infection intensities during the early stages of infection, i.e. before the parasite reaches the adult stage, were similar in both eel species in the laboratory, suggesting that the different susceptibilities do not manifest until the later infection stages. Despite the similarity of infection intensity, gene expression changes upon *A. crassus* infection were more pronounced and more diverse in the European eel. This occurred in both head kidney and spleen and at both time points examined, providing evidence for the involvement of both organs to an immune response and its continuous activation. Gene expression in the Japanese eel indicated that an immune response during these early infection stages is only of minor importance for coping with *A. crassus*. This difference between host species suggests that an ineffective and prolonged immune response early during infection in the European eel may contribute to the higher susceptibility previously reported for this species. These results add to observations in other novel host-parasite systems that increased immune gene expression is associated with susceptibility rather than increased resistance. The results reported here also support the hypothesis that susceptible novel hosts lack effective defence mechanisms to newly acquired parasites. However, the mechanisms that lead to higher resistance in the Japanese eel remain elusive. The induction of ineffective defence strategies may have caused fitness costs in the form of disruption of energetic and physiological processes that were observed here for the European eel but not for the Japanese eel. Thus, physiological and metabolic processes may contribute substantially to the higher susceptibility and impaired health of the European eel. Integrating these processes may improve our understanding of differing host responses and susceptibilities to parasites in general and the negative impact of invasive parasites in particular.

Unlike the picture drawn by experimental infections of European eels in this and previous studies, the examination of naturally infected European eels here suggested that coping mechanisms are evolving. Encapsulation of *A. crassus* by the European eel was associated with a reduced *A. crassus* load, particularly of adults, and this relationship was temporally stable. The reduced load and the fact that encapsulation was previously reported for the Japanese eel in both natural and experimental infections indicate that encapsulation may constitute a sign of ongoing adaptation of the European eel to its novel parasite. This was also reflected in a reduced activation of the adaptive immune system with increasing number of capsules in responding individuals as indicated by their negative correlation with *mhc II* gene expression in this study. However, some native parasites had higher abundances in

European eel individuals responding to *A. crassus* compared to non-responders. Hence, adapting to *A. crassus* may come at the expense of coping less efficiently with native parasites. This higher abundance of native parasites may explain why both immune and metabolic gene expression did not differ between responding and non-responding individuals despite the markedly reduced abundance of costly adult *A. crassus* stages. Furthermore, the observation of site-specific encapsulation prevalence suggests that additional environmental factors may determine the response phenotype.

References

Abbas, A. K., Lichtman, A. H. & Pillai, S. 2015. *Cellular and molecular immunology*, 8th ed. Elsevier Inc., Philadelphia, USA.

Abbas, A. K., Lichtman, A. H. & Pillai, S. 2018. *Cellular and molecular immunology*, 9th ed. Elsevier Inc., Philadelphia, USA.

Abbate, J. L., Ezenwa, V. O., Guégan, J.-F., Choisy, M., Nacher, M. & Roche, B. 2018. Disentangling complex parasite interactions: protection against cerebral malaria by one helminth species is jeopardized by co-infection with another. *PLOS Neglected Tropical Diseases* **12**: e0006483.

Adlard, R. D., Miller, T. L. & Smit, N. J. 2015. The butterfly effect: parasite diversity, environment, and emerging disease in aquatic wildlife. *Trends in Parasitology* **31**: 160-166.

Aieta, A. E. & Oliveira, K. 2009. Distribution, prevalence, and intensity of the swim bladder parasite *Anguillicola crassus* in New England and eastern Canada. *Diseases of Aquatic Organisms* **84**: 229-235.

Alderman, D. J., Polglase, J. L. & Frayling, M. 1987. *Aphanomyces astaci* pathogenicity under laboratory and field conditions. *Journal of Fish Diseases* **10**: 385-393.

Als, T. D., Hansen, M. M., Maes, G. E., Castonguay, M., Riemann, L., Aarestrup, K., Munk, P., Sparholt, H., Hanel, R. & Bernatchez, L. 2011. All roads lead to home: panmixia of European eel in the Sargasso Sea. *Molecular Ecology* **20**: 1333-1346.

Alvarez Rojas, C. A., Ansell, B. R. E., Hall, R. S., Gasser, R. B., Young, N. D., Jex, A. R. & Scheerlinck, J. P. Y. 2015. Transcriptional analysis identifies key genes involved in metabolism, fibrosis/tissue repair and the immune response against *Fasciola hepatica* in sheep liver. *Parasites & Vectors* **8**: 14.

Alvarez-Pellitero, P. 2008. Fish immunity and parasite infections: from innate immunity to immunoprophylactic prospects. *Veterinary Immunology and Immunopathology* **126**: 171-198.

Anderson, R. M. & May, R. M. 1982. Coevolution of hosts and parasites. *Parasitology* **85**: 411-426.

Aoyama, J., Watanabe, S., Miller, M. J., Mochioka, N., Otake, T., Yoshinaga, T. & Tsukamoto, K. 2014. Spawning sites of the Japanese eel in relation to oceanographic structure and the West Mariana Ridge. *Plos One* **9**.

Atkinson, C. T., Saili, K. S., Utzurrum, R. B. & Jarvi, S. I. 2013. Experimental evidence for evolved tolerance to avian malaria in a wild population of low elevation Hawai'i 'amakihi (*Hemignathus virens*). *Ecohealth* **10**: 366-375.

Audenaert, V., Huyse, T., Goemans, G., Belpaire, C. & Volckaert, F. A. M. 2003. Spatio-temporal dynamics of the parasitic nematode *Anguillicola crassus* in Flanders, Belgium. *Diseases of Aquatic Organisms* **56**: 223-233.

Aufauvre, J., Misme-Aucouturier, B., Vigues, B., Texier, C., Delbac, F. & Blot, N. 2014. Transcriptome analyses of the honeybee response to *Nosema ceranae* and insecticides. *Plos One* **9**: 12.

Augert, A., Payre, C., de Launoit, Y., Gil, J., Lambeau, G. & Bernard, D. 2009. The M-type receptor PLA2R regulates senescence through the p53 pathway. *Embo Reports* **10**: 271-277.

Babayan, S. A., Liu, W., Hamilton, G., Kilbride, E., Rynkiewicz, E. C., Clerc, M. & Pedersen, A. B. 2018. The immune and non-immune pathways that drive chronic gastrointestinal helminth burdens in the wild. *Frontiers in Immunology* **9**: 12.

Baerwald, M. R. 2013. Temporal expression patterns of rainbow trout immune-related genes in response to *Myxobolus cerebralis* exposure. *Fish & Shellfish Immunology* **35**: 965-971.

Baltazar-Soares, M., Biastoch, A., Harrod, C., Hanel, R., Marohn, L., Prigge, E., Evans, D., Bodles, K., Behrens, E., Boning, C. W. & Eizaguirre, C. 2014. Recruitment collapse and population structure of the European eel shaped by local ocean current dynamics. *Current Biology* **24**: 104-108.

Barry, J., McLeish, J., Dodd, J. A., Turnbull, J. F., Boylan, P. & Adams, C. E. 2014. Introduced parasite *Anguillicola crassus* infection significantly impedes swim bladder function in the European eel *Anguilla anguilla* (L.). *Journal of Fish Diseases* **37**: 921-924.

Barry, J., Newton, M., Dodd, J. A., Evans, D., Newton, J. & Adams, C. E. 2017. The effect of foraging and ontogeny on the prevalence and intensity of the invasive parasite *Anguillicola crassus* in the European eel *Anguilla anguilla*. *Journal of Fish Diseases* **40**: 1213-1222.

Barse, A. M. & Secor, D. H. 1999. An exotic nematode parasite of the American eel. *Fisheries* **24**: 6-10.

Becerra-Jurado, G., Cruikshanks, R., O'Leary, C., Kelly, F., Poole, R. & Gargan, P. 2014. Distribution, prevalence and intensity of *Anguillicola crassus* (Nematoda) infection in *Anguilla anguilla* in the Republic of Ireland. *Journal of Fish Biology* **84**: 1046-1062.

Behringer, D. C., Karvonen, A. & Bojko, J. 2018. Parasite avoidance behaviours in aquatic environments. *Philosophical Transactions of the Royal Society B-Biological Sciences* **373**: 19.

Benesh, D. P. & Kalbe, M. 2016. Experimental parasite community ecology: intraspecific variation in a large tapeworm affects community assembly. *Journal of Animal Ecology* **85**: 1004-1013.

Benjamini, Y. & Hochberg, Y. 1995. Controlling the false discovery rate - A practical and powerful approach to multiple testing. *Journal of the Royal Statistical Society Series B-Methodological* **57**: 289-300.

Bergersen, E. P. & Anderson, D. E. 1997. The distribution and spread of *Myxobolus cerebralis* in the United States. *Fisheries* **22**: 6-7.

Bertolino, S. 2009. Animal trade and non-indigenous species introduction: the world-wide spread of squirrels. *Diversity and Distributions* **15**: 701-708.

Bertolino, S., di Montezemolo, N. C., Preatoni, D. G., Wauters, L. A. & Martinoli, A. 2014. A grey future for Europe: *Sciurus carolinensis* is replacing native red squirrels in Italy. *Biological Invasions* **16**: 53-62.

Betts, A., Rafaluk, C. & King, K. C. 2016. Host and parasite evolution in a tangled bank. *Trends in Parasitology* **32**: 863-873.

Blanc, G., Bonneau, S., Biagianti, S. & Petter, A. J. 1992. Description of the larval stages of *Anguillicola crassus* (Nematoda, Dracunculoidea) using light and scanning electron microscopy. *Aquatic Living Resources* **5**: 307-318.

Bonneau, S., Blanc, G. & Petter, A. J. 1991. Studies on the biology of the early larval stages of *Anguillicola crassus* (Dracunculoidea, Nematoda) - specificity of the intermediate host and effect of temperature on rate of development. *Bulletin Francais De La Peche Et De La Pisciculture*: 1-6.

Bonneaud, C., Balenger, S. L., Russell, A. F., Zhang, J. W., Hill, G. E. & Edwards, S. V. 2011. Rapid evolution of disease resistance is accompanied by functional changes in gene expression in a wild bird. *Proceedings of the National Academy of Sciences of the United States of America* **108**: 7866-7871.

Bonneaud, C., Balenger, S. L., Zhang, J. W., Edwards, S. V. & Hill, G. E. 2012. Innate immunity and the evolution of resistance to an emerging infectious disease in a wild bird. *Molecular Ecology* **21**: 2628-2639.

Bonneaud, C., Perez-Tris, J., Federici, P., Chastel, O. & Sorci, G. 2006. Major histocompatibility alleles associated with local resistance to malaria in a passerine. *Evolution* **60**: 383-389.

Bornarel, V., Lambert, P., Briand, C., Antunes, C., Belpaire, C., Ciccotti, E., Diaz, E., Diserud, O., Doherty, D., Domingos, I., Evans, D., de Graaf, M., O'Leary, C., Pedersen, M., Poole, R., Walker, A., Wickström, H., Beaulaton, L. & Drouineau, H. 2017. Modelling the recruitment of European eel (*Anguilla anguilla*) throughout its European range. *ICES Journal of Marine Science* **75**: 541-552.

Bracamonte, S. E., Baltazar-Soares, M. & Eizaguirre, C. 2015. Characterization of MHC class II genes in the critically endangered European eel (*Anguilla anguilla*). *Conservation Genetics Resources* **7**: 859-870.

Bracamonte, S. E., Johnston, P. R., Knopf, K. & Monaghan, M. T. 2019. Experimental infection with *Anguillicola crassus* alters immune gene expression in both spleen and head kidney of the European eel (*Anguilla anguilla*). *Marine Genomics* **45**: 28-37.

Bracamonte, S. E., Johnston, P. R., Monaghan, M. T. & Knopf, K. in press. Gene expression response to a nematode parasite in novel and native eel hosts. *Ecology and Evolution*.

Britton, J. R. & Orsi, M. L. 2012. Non-native fish in aquaculture and sport fishing in Brazil: economic benefits versus risks to fish diversity in the upper River Parana Basin. *Reviews in Fish Biology and Fisheries* **22**: 555-565.

Bronte, V. & Pittet, M. J. 2013. The spleen in local and systemic regulation of immunity. *Immunity* **39**: 806-818.

Brown, J. H., Jardetzky, T. S., Gorga, J. C., Stern, L. J., Urban, R. G., Strominger, J. L. & Wiley, D. C. 1993. 3-dimensional structure of the human class-II histocompatibility antigen HLA-DR1. *Nature* **364**: 33-39.

Brown, K. L. & Hancock, R. E. W. 2006. Cationic host defense (antimicrobial) peptides. *Current Opinion in Immunology* **18**: 24-30.

Brown, M., Hablützel, P., Friberg, I. M., Thomason, A. G., Stewart, A., Pachebat, J. A. & Jackson, J. A. 2016. Seasonal immunoregulation in a naturally-occurring vertebrate. *BMC Genomics* **17**: 18.

Buchmann, K. 2012. Fish immune responses against endoparasitic nematodes - experimental models. *Journal of Fish Diseases* **35**: 623-635.

Cadi, A. & Joly, P. 2004. Impact of the introduction of the red-eared slider (*Trachemys scripta elegans*) on survival rates of the European pond turtle (*Emys orbicularis*). *Biodiversity and Conservation* **13**: 2511-2518.

Cerda, J. & Finn, R. N. 2010. Piscine aquaporins: an overview of recent advances. *Journal of Experimental Zoology Part A - Ecological Genetics and Physiology* **313A**: 623-650.

Chucholl, C. 2013. Invaders for sale: trade and determinants of introduction of ornamental freshwater crayfish. *Biological Invasions* **15**: 125-141.

Clarke, K. R. 1993. Nonparametric multivariate analyses of changes in community structure. *Australian Journal of Ecology* **18**: 117-143.

Coppe, A., Pujolar, J. M., Maes, G. E., Larsen, P. F., Hansen, M. M., Bernatchez, L., Zane, L. & Bortoluzzi, S. 2010. Sequencing, de novo annotation and analysis of the first *Anguilla anguilla* transcriptome: EeelBase opens new perspectives for the study of the critically endangered European eel. *BMC Genomics* **11**: 635.

Crotty, S. 2015. A brief history of T cell help to B cells. *Nature Reviews Immunology* **15**: 185-189.

Cunningham, A. A. 1996. Disease risks of wildlife translocations. *Conservation Biology* **10**: 349-353.

Cézilly, F., Perrot-Minnot, M. J. & Rigaud, T. 2014. Cooperation and conflict in host manipulation: interactions among macro-parasites and micro-organisms. *Frontiers in Microbiology* **5**: 10.

Dangel, K. C., Keppel, M., Le, T. T. Y., Grabner, D. & Sures, B. 2015. Competing invaders: performance of two *Anguillicola* species in Lake Bracciano. *International Journal for Parasitology-Parasites and Wildlife* **4**: 119-124.

Dangel, K. C., Keppel, M. & Sures, B. 2013. Can differences in life cycle explain differences in invasiveness? - A study on *Anguillicola novaezelandiae* in the European eel. *Parasitology* **140**: 1831-1836.

Daszak, P., Cunningham, A. A. & Hyatt, A. D. 2000. Wildlife ecology - Emerging infectious diseases of wildlife - threats to biodiversity and human health. *Science* **287**: 443-449.

Daszak, P., Cunningham, A. A. & Hyatt, A. D. 2001. Anthropogenic environmental change and the emergence of infectious diseases in wildlife. *Acta Tropica* **78**: 103-116.

Davy, C. M., Donaldson, M. E., Willis, C. K. R., Saville, B. J., McGuire, L. P., Mayberry, H., Wilcox, A., Wibbelt, G., Misra, V., Bollinger, T. & Kyle, C. J. 2017. The other white-nose syndrome transcriptome: tolerant and susceptible hosts respond differently to the pathogen *Pseudogymnoascus destructans*. *Ecology and Evolution* **7**: 7161-7170.

De Charleroy, D., Grisez, L., Thomas, K., Belpaire, C. & Ollevier, F. 1990. The life-cycle of *Anguillicola crassus*. *Diseases of Aquatic Organisms* **8**: 77-84.

Dekker, W. 2003. Did lack of spawners cause the collapse of the European eel, *Anguilla anguilla*? *Fisheries Management and Ecology* **10**: 365-376.

Demann, F. & Wegner, K. M. 2019. Infection by invasive parasites increases susceptibility of native hosts to secondary infection via modulation of cellular immunity. *Journal of Animal Ecology* **88**: 427-438.

Densmore, C. L., Ottinger, C. A., Blazer, V. S. & Iwanowicz, L. R. 2004. Immunomodulation and disease resistance in postyearling rainbow trout infected with *Myxobolus cerebralis*, the causative agent of Whirling disease. *Journal of Aquatic Animal Health* **16**: 73-82.

Dezfuli, B. S., Bo, T., Lorenzoni, M., Shinn, A. P. & Giari, L. 2015. Fine structure and cellular responses at the host-parasite interface in a range of fish-helminth systems. *Veterinary Parasitology* **208**: 272-279.

Dezfuli, B. S., Giari, L., Castaldelli, G., Lanzoni, M., Rossi, R., Lorenzoni, M. & Kennedy, C. R. 2014. Temporal and spatial changes in the composition and structure of helminth component communities in European eels *Anguilla anguilla* in an Adriatic coastal lagoon and some freshwaters in Italy. *Parasitology Research* **113**: 113-120.

Diekmann, M., Simon, J. & Salva, J. 2019. On the actual recruitment of European eel (*Anguilla anguilla*) in the River Ems, Germany. *Fisheries Management and Ecology* **26**: 20-30.

Dionne, M., Miller, K. M., Dodson, J. J. & Bernatchez, L. 2009. MHC standing genetic variation and pathogen resistance in wild Atlantic salmon. *Philosophical Transactions of the Royal Society B-Biological Sciences* **364**: 1555-1565.

Dittel, A. I. & Epifanio, C. E. 2009. Invasion biology of the Chinese mitten crab *Eriochier sinensis*: a brief review. *Journal of Experimental Marine Biology and Ecology* **374**: 79-92.

Dobson, A. & Foufopoulos, J. 2001. Emerging infectious pathogens of wildlife. *Philosophical Transactions of the Royal Society of London Series B-Biological Sciences* **356**: 1001-1012.

Dobson, A., Lafferty, K. D., Kuris, A. M., Hechinger, R. F. & Jetz, W. 2008. Homage to Linnaeus: How many parasites? How many hosts? *Proceedings of the National Academy of Sciences of the United States of America* **105**: 11482-11489.

Drouineau, H., Durif, C., Castonguay, M., Mateo, M., Rochard, E., Verreault, G., Yokouchi, K. & Lambert, P. 2018. Freshwater eels: a symbol of the effects of global change. *Fish and Fisheries* **19**: 903-930.

Duffy, M. A. & Sivars-Becker, L. 2007. Rapid evolution and ecological host-parasite dynamics. *Ecology Letters* **10**: 44-53.

Duncan, A. B. & Little, T. J. 2007. Parasite-driven genetic change in a natural population of *Daphnia*. *Evolution* **61**: 796-803.

Dunn, A. M. & Hatcher, M. J. 2015. Parasites and biological invasions: parallels, interactions, and control. *Trends in Parasitology* **31**: 189-199.

Durif, C., Dufour, S. & Elie, P. 2005. The silvering process of *Anguilla anguilla*: a new classification from the yellow resident to the silver migrating stage. *Journal of Fish Biology* **66**: 1025-1043.

Díaz, A. & Allen, J. E. 2007. Mapping immune response profiles: The emerging scenario from helminth immunology. *European Journal of Immunology* **37**: 3319-3326.

Egusa, S. 1979. Notes on the culture of the European eel (*Anguilla anguilla* L.) in Japanese eel-farming ponds. *Rapports et Procès-Verbaux des Réunions, Conseil International pour l'Exploration de la Mer* **174**: 51-58.

Ehret, T., Spork, S., Dieterich, C., Lucius, R. & Heitlinger, E. 2017. Dual RNA-seq reveals no plastic transcriptional response of the coccidian parasite *Eimeria falciformis* to host immune defenses. *BMC Genomics* **18**: 17.

Eizaguirre, C., Lenz, T. L., Kalbe, M. & Milinski, M. 2012a. Divergent selection on locally adapted major histocompatibility complex immune genes experimentally proven in the field. *Ecology Letters* **15**: 723-731.

Eizaguirre, C., Lenz, T. L., Kalbe, M. & Milinski, M. 2012b. Rapid and adaptive evolution of MHC genes under parasite selection in experimental vertebrate populations. *Nature Communications* **3**: 621.

Eizaguirre, C., Lenz, T. L., Sommerfeld, R. D., Harrod, C., Kalbe, M. & Milinski, M. 2011. Parasite diversity, patterns of MHC II variation and olfactory based mate choice in diverging three-spined stickleback ecotypes. *Evolutionary Ecology* **25**: 605-622.

El-Shehabi, F., Marcogliese, D. J. & Oliveira, K. 2018. New North American paratenic hosts of *Anguillicola crassus* and molecularly inferred source of invasion. *Aquatic Invasions* **13**: 231-246.

Elgueta, R., Benson, M. J., de Vries, V. C., Wasiuk, A., Guo, Y. X. & Noelle, R. J. 2009. Molecular mechanism and function of CD40/CD40L engagement in the immune system. *Immunological Reviews* **229**: 152-172.

Elliott, S., Sinclair, A., Collins, H., Rice, L. & Jelkmann, W. 2014. Progress in detecting cell-surface protein receptors: the erythropoietin receptor example. *Annals of Hematology* **93**: 181-192.

Ellis, A. E. 1980. Antigen-trapping in the spleen and kidney of the plaice *Pleuronectes platessa* L. *Journal of Fish Diseases* **3**: 413-426.

Ellison, A. R., Tunstall, T., DiRenzo, G. V., Hughey, M. C., Rebollar, E. A., Belden, L. K., Harris, R. N., Ibanez, R., Lips, K. R. & Zamudio, K. R. 2015. More than skin deep: functional genomic basis for resistance to amphibian chytridiomycosis. *Genome Biology and Evolution* **7**: 286-298.

Emms, D. M. & Kelly, S. 2015. OrthoFinder: solving fundamental biases in whole genome comparisons dramatically improves orthogroup inference accuracy. *Genome Biology* **16**.

Ercan, D., Andreou, D., Sana, S., Öntas, C., Baba, E., Top, N., Karakus, U., Tarkan, A. S. & Gozlan, R. E. 2015. Evidence of threat to European economy and biodiversity following the introduction of an alien pathogen on the fungal-animal boundary. *Emerging Microbes & Infections* **4**: 6.

Eskew, E. A., Shock, B. C., LaDouceur, E. E. B., Keel, K., Miller, M. R., Foley, J. E. & Todd, B. D. 2018. Gene expression differs in susceptible and resistant amphibians exposed to *Batrachochytrium dendrobatidis*. *Royal Society Open Science* **5**: 23.

Evans, T. G. 2015. Considerations for the use of transcriptomics in identifying the 'genes that matter' for environmental adaptation. *Journal of Experimental Biology* **218**: 1925-1935.

Fabriek, B. O., Dijkstra, C. D. & van den Berg, T. K. 2005. The macrophage scavenger receptor CD163. *Immunobiology* **210**: 153-160.

Falcon, S. & Gentleman, R. 2007. Using GOstats to test gene lists for GO term association. *Bioinformatics* **23**: 257-258.

Fazio, G., Mone, H., Da Silva, C., Simon-Levert, G., Allienne, J. F., Lecomte-Finiger, R. & Sasal, P. 2009. Changes in gene expression in European eels (*Anguilla anguilla*) induced by infection with swim bladder nematodes (*Anguillicola crassus*). *Journal of Parasitology* **95**: 808-816.

Fazio, G., Mone, H., Lecomte-Finiger, R. & Sasal, R. 2008a. Differential gene expression analysis in European eels (*Anguilla Anguilla*, L. 1758) naturally infected by macroparasites. *Journal of Parasitology* **94**: 571-577.

Fazio, G., Sasal, P., Lecomte-Finiger, R., Da Silva, C., Fumet, B. & Mone, H. 2008b. Macroparasite communities in European eels, *Anguilla anguilla*, from French Mediterranean lagoons, with special reference to the invasive species *Anguillicola crassus* and *Pseudodactylogyrus* spp. *Knowledge and Management of Aquatic Ecosystems*: 12.

Fazio, G., Sasal, P., Mouahid, G., Lecomte-Finiger, R. & Mone, H. 2012. Swim bladder nematiodes (*Anguillicoloides crassus*) disturb silvering in European eels (*Anguilla anguilla*). *Journal of Parasitology* **98**: 695-705.

Feis, M. E., Goedknegt, M. A., Thieltges, D. W., Buschbaum, C. & Wegner, K. M. 2016. Biological invasions and host-parasite coevolution: different coevolutionary trajectories along separate parasite invasion fronts. *Zoology* **119**: 366-374.

Feis, M. E., John, U., Lokmer, A., Luttikhuizen, P. C. & Wegner, K. M. 2018. Dual transcriptomics reveals co-evolutionary mechanisms of intestinal parasite infections in blue mussels *Mytilus edulis*. *Molecular Ecology* **27**: 1505-1519.

Field, K. A., Johnson, J. S., Lilley, T. M., Reeder, S. M., Rogers, E. J., Behr, M. J. & Reeder, D. M. 2015. The white-nose syndrome transcriptome: activation of anti-fungal host responses in wing tissue of hibernating little brown myotis. *Plos Pathogens* **11**: 29.

Frick, W. F., Pollock, J. F., Hicks, A. C., Langwig, K. E., Reynolds, D. S., Turner, G. G., Butchkoski, C. M. & Kunz, T. H. 2010. An emerging disease causes regional population collapse of a common North American bat species. *Science* **329**: 679-682.

Friedland, K. D., Miller, M. J. & Knights, B. 2007. Oceanic changes in the Sargasso Sea and declines in recruitment of the European eel. *ICES Journal of Marine Science* **64**: 519-530.

Froese, R. & Pauly, D. (2019) FishBase. Vol. 2019. pp.

Fèvre, E. M., Bronsvoort, B., Hamilton, K. A. & Cleaveland, S. 2006. Animal movements and the spread of infectious diseases. *Trends in Microbiology* **14**: 125-131.

Gause, W. C., Wynn, T. A. & Allen, J. E. 2013. Type 2 immunity and wound healing: evolutionary refinement of adaptive immunity by helminths. *Nature Reviews Immunology* **13**: 607-614.

Geeraerts, C. & Belpaire, C. 2010. The effects of contaminants in European eel: a review. *Ecotoxicology* **19**: 239-266.

Gibson, A. K., Jokela, J. & Lively, C. M. 2016. Fine-scale spatial covariation between infection prevalence and susceptibility in a natural population. *American Naturalist* **188**: 1-14.

Glasauer, S. M. K. & Neuhauss, S. C. F. 2014. Whole-genome duplication in teleost fishes and its evolutionary consequences. *Molecular Genetics and Genomics* **289**: 1045-1060.

Goedknegt, M. A., Feis, M. E., Wegner, K. M., Luttikhuizen, P. C., Buschbaum, C., Camphuysen, K., van der Meer, J. & Thieltges, D. W. 2016. Parasites and marine invasions: Ecological and evolutionary perspectives. *Journal of Sea Research* **113**: 11-27.

Gollock, M. J., Kennedy, C. R. & Brown, J. A. 2005. European eels, *Anguilla anguilla* (L.), infected with *Anguillicola crassus* exhibit a more pronounced stress response to severe hypoxia than uninfected eels. *Journal of Fish Diseases* **28**: 429-436.

Gozlan, R. E., St-Hilaire, S., Feist, S. W., Martin, P. & Kent, M. L. 2005. Biodiversity - Disease threat to European fish. *Nature* **435**: 1046-1046.

Grabherr, M. G., Haas, B. J., Yassour, M., Levin, J. Z., Thompson, D. A., Amit, I., Adiconis, X., Fan, L., Raychowdhury, R., Zeng, Q. D., Chen, Z. H., Mauceli, E., Hacohen, N., Gnirke, A., Rhind, N., di Palma, F., Birren, B. W., Nusbaum, C., Lindblad-Toh, K., Friedman, N. & Regev, A. 2011. Full-length transcriptome assembly from RNA-Seq data without a reference genome. *Nature Biotechnology* **29**: 644-U130.

Graham, A. L. 2008. Ecological rules governing helminth-microparasite coinfection. *Proceedings of the National Academy of Sciences of the United States of America* **105**: 566-570.

Graham, A. L., Allen, J. E. & Read, A. F. 2005. Evolutionary causes and consequences of immunopathology. *Annual Review of Ecology Evolution and Systematics* **36**: 373-397.

Granata, F., Petraroli, A., Boilard, E., Bezzine, S., Bollinger, J., Del Vecchio, L., Gelb, M. H., Lambeau, G., Marone, G. & Triggiani, M. 2005. Activation of cytokine production by secreted phospholipase A_2 in human lung macrophages expressing the M-type receptor. *Journal of Immunology* **174**: 464-474.

Grogan, L. F., Cashins, S. D., Skerratt, L. F., Berger, L., McFadden, M. S., Harlow, P., Hunter, D. A., Scheele, B. & Mulvenna, J. 2018. Evolution of resistance to chytridiomycosis is associated with a robust early immune response. *Molecular Ecology* **27**: 919-934.

Guo, Z. Y., Gonzalez, J. F., Hernandez, J. N., McNeilly, T. N., Corripio-Miyar, Y., Frew, D., Morrison, T., Yu, P. & Li, R. W. 2016. Possible mechanisms of host resistance to *Haemonchus contortus* infection in sheep breeds native to the Canary Islands. *Scientific Reports* **6**: 14.

Gururajan, M., Simmons, A., Dasu, T., Spear, B. T., Calulot, C., Robertson, D. A., Wiest, D. L., Monroe, J. G. & Bondada, S. 2008. Early growth response genes regulate B cell development, proliferation, and immune response. *Journal of Immunology* **181**: 4590-4602.

Gérard, C., Trancart, T., Amilhat, E., Faliex, E., Virag, L., Feunteun, E. & Acou, A. 2013. Influence of introduced vs. native parasites on the body condition of migrant silver eels. *Parasite* **20**: 10.

Haas, B. J., Papanicolaou, A., Yassour, M., Grabherr, M., Blood, P. D., Bowden, J., Couger, M. B., Eccles, D., Li, B., Lieber, M., MacManes, M. D., Ott, M., Orvis, J., Pochet, N., Strozzi, F., Weeks, N., Westerman, R., William, T., Dewey, C. N., Henschel, R., Leduc, R. D., Friedman, N. & Regev, A. 2013. De novo transcript sequence reconstruction from RNA-seq using the Trinity platform for reference generation and analysis. *Nature Protocols* **8**: 1494-1512.

Haase, D., Rieger, J. K., Witten, A., Stoll, M., Bornberg-Bauer, E., Kalbe, M., Schmidt-Drewello, A., Scharsack, J. P. & Reusch, T. B. H. 2016. Comparative transcriptomics of stickleback immune gene responses upon infection by two helminth parasites, *Diplostomum pseudospathaceum* and *Schistocephalus solidus*. *Zoology* **119**: 307-313.

Haenen, O. L. M., Grisez, L., Decharleroy, D., Belpaire, C. & Ollevier, F. 1989. Experimentally induced infections of European eel *Anguilla anguilla* with *Anguillicola crassus* (Nematoda, Dracunculoidea) and subsequent migration of larvae. *Diseases of Aquatic Organisms* **7**: 97-101.

Haenen, O. L. M. & van Banning, P. 1990. Detection of larvae of *Anguillicola crassus* (an eel swimbladder nematode) in freshwater fish species. *Aquaculture* **87**: 103-109.

Haenen, O. L. M., Vanwijngaarden, T. A. M. & Borgsteede, F. H. M. 1994. An improved method for the production of infective 3rd-stage juveniles of *Anguillicola crassus*. *Aquaculture* **123**: 163-165.

Hafer, N. & Milinski, M. 2016. Inter- and intraspecific conflicts between parasites over host manipulation. *Proceedings of the Royal Society B-Biological Sciences* **283**: 9.

Han, Y. S., Chang, Y. T., Taraschewski, H., Chang, S. L., Chen, C. C. & Tzeng, W. N. 2008. The swimbladder parasite *Anguillicola crassus* in native Japanese eels and exotic American eels in Taiwan. *Zoological Studies* **47**: 667-675.

Hatcher, M. J., Dick, J. T. A. & Dunn, A. M. 2012. Diverse effects of parasites in ecosystems: linking interdependent processes. *Frontiers in Ecology and the Environment* **10**: 186-194.

Hedrick, R. P., El-Matbouli, M., Adkison, M. A. & MacConnell, E. 1998. Whirling disease: re-emergence among wild trout. *Immunological Reviews* **166**: 365-376.

Hein, J. L., Arnott, S. A., Roumillat, W. A., Allen, D. M. & de Buron, I. 2014. Invasive swimbladder parasite *Anguillicoloides crassus*: infection status 15 years after discovery in wild populations of American eel *Anguilla rostrata*. *Diseases of Aquatic Organisms* **107**: 199-209.

Heitlinger, E., Taraschewski, H., Weclawski, U., Gharbi, K. & Blaxter, M. 2014. Transcriptome analyses of *Anguillicola crassus* from native and novel hosts. *PeerJ* **2**: 22.

Heitlinger, E. G., Laetsch, D. R., Weclawski, U., Han, Y. S. & Taraschewski, H. 2009. Massive encapsulation of larval *Anguillicoloides crassus* in the intestinal wall of Japanese eels. *Parasites & Vectors* **2**: 11.

Henkel, C. (2019) Replication data for: Rapid *de novo* assembly of the European eel genome from nanopore sequencing reads. (Norwegian University of Life, S., ed.). pp. DataverseNO.

Henkel, C. V., Burgerhout, E., de Wijze, D. L., Dirks, R. P., Minegishi, Y., Jansen, H. J., Spaink, H. P., Dufour, S., Weltzien, F. A., Tsukamoto, K. & van den Thillart, G. 2012a. Primitive duplicate Hox clusters in the European eel's genome. *Plos One* **7**: 9.

Henkel, C. V., Dirks, R. P., de Wijze, D. L., Minegishi, Y., Aoyama, J., Jansen, H. J., Turner, B., Knudsen, B., Bundgaard, M., Hvam, K. L., Boetzer, M., Pirovano, W., Weltzien, F. A., Dufour, S., Tsukamoto, K., Spaink, H. P. & van den Thillart, G. 2012b. First draft genome sequence of the Japanese eel, *Anguilla japonica*. *Gene* **511**: 195-201.

Holdich, D. M., Reynolds, J. D., Souty-Grosset, C. & Sibley, P. J. 2009. A review of the ever increasing threat to European crayfish from non-indigenous crayfish species. *Knowledge and Management of Aquatic Ecosystems*: 46.

Hosler, J. P., Ferguson-Miller, S. & Mills, D. A. (2006) Energy transduction: proton transfer through the respiratory complexes. In: *Annual Review of Biochemistry*, Vol. 75. pp. 165-187 Annual Review of Biochemistry. Annual Reviews, Palo Alto.

Hothorn, T., Bretz, F. & Westfall, P. 2008. Simultaneous inference in general parametric models. *Biometrical Journal* **50**: 346-363.

Hu, Z.-L., Bao, J. & Reecy, J. M. 2008. CateGOrizer: a web-based program to batch analyse gene ontology classification categories. *Online Journal of Bioinformatics* **9**: 108-112.

Huang, Y., Chain, F. J. J., Panchal, M., Eizaguirre, C., Kalbe, M., Lenz, T. L., Samonte, I. E., Stoll, M., Bornberg-Bauer, E., Reusch, T. B. H., Milinski, M. & Feulner, P. G. D. 2016. Transcriptome profiling of immune tissues reveals habitat-specific gene expression between lake and river sticklebacks. *Molecular Ecology* **25**: 943-958.

Hulme, P. E., Bacher, S., Kenis, M., Klotz, S., Kühn, I., Minchin, D., Nentwig, W., Olenin, S., Panov, V., Pergl, J., Pysek, P., Roques, A., Sol, D., Solarz, W. & Vilà, M. 2008. Grasping at the routes of biological invasions: a framework for integrating pathways into policy. *Journal of Applied Ecology* **45**: 403-414.

Höglund, J., Andersson, J. & Hardig, J. 1992. Hematological responses in the European eel, *Anguilla anguilla* L, to sublethal infestation by *Anguillicola crassus* in a thermal effluent of the Swedish baltic. *Journal of Fish Diseases* **15**: 507-514.

ICES (2012) Report of the 2012 Session of the Joint EIFAAC/ICES Working Group on Eels. pp. 824. ICES CM 2012/ACOM:18, Copenhagen, Denmark.

ICES (2018) Report of the Joint EIFAAC/ICES/GFCM Working Group on Eels (WGEEL). pp. 152. ICES CM 2018/ACOM:15, Gdańsk, Poland.

Iglesias, R., García-Estévez, J. M., Ayres, C., Acuña, A. & Cordero-Rivera, A. 2015. First reported outbreak of severe spirorchiidiasis in *Emys orbicularis*, probably resulting from a parasite spillover event. *Diseases of Aquatic Organisms* **113**: 75-80.

IUCN (2018) The IUCN Red List of threatened species. Vol. 2019. pp.

Jacoby, D. M. P., Casselman, J. M., Crook, V., DeLucia, M.-B., Ahn, H., Kaifu, K., Kurwie, T., Sasal, P., Silfvergrip, A. M. C., Smith, K. G., Uchida, K., Walker, A. M. & Gollock, M. J. 2015. Synergistic patterns of threat and the challenges facing global anguillid eel conservation. *Global Ecology and Conservation* **4**: 321-333.

Jakob, E., Hanel, R., Klimpel, S. & Zumholz, K. 2009. Salinity dependence of parasite infestation in the European eel *Anguilla anguilla* in northern Germany. *ICES Journal of Marine Science* **66**: 358-366.

Jakob, E., Walter, T. & Hanel, R. 2016. A checklist of the protozoan and metazoan parasites of European eel (*Anguilla anguilla*): checklist of *Anguilla anguilla* parasites. *Journal of Applied Ichthyology* **32**: 757-804.

Jansen, H. J., Liem, M., Jong-Raadsen, S. A., Dufour, S., Weltzien, F. A., Swinkels, W., Koelewijn, A., Palstra, A. P., Pelster, B., Spaink, H. P., van den Thillart, G. E., Dirks, R. P. & Henkel, C. V. 2017. Rapid *de novo* assembly of the European eel genome from nanopore sequencing reads. *Scientific Reports* **7**: 13.

Jenkins, S. R., Perry, B. D. & Winkler, W. G. 1988. Ecology and epidemiology of raccoon rabies. *Reviews of Infectious Diseases* **10**: S620-S625.

Johnson, P. T. J. & Hoverman, J. T. 2012. Parasite diversity and coinfection determine pathogen infection success and host fitness. *Proceedings of the National Academy of Sciences* **109**: 9006-9011.

Johnson, P. T. J. & Paull, S. H. 2011. The ecology and emergence of diseases in fresh waters. *Freshwater Biology* **56**: 638-657.

Kalbe, M., Eizaguirre, C., Scharsack, J. P. & Jakobsen, P. J. 2016. Reciprocal cross infection of sticklebacks with the diphyllobothriidean cestode *Schistocephalus solidus* reveals consistent population differences in parasite growth and host resistance. *Parasites & Vectors* **9**: 12.

Kalbe, M. & Kurtz, J. 2006. Local differences in immunocompetence reflect resistance of sticklebacks against the eye fluke *Diplostomum pseudospathaceum*. *Parasitology* **132**: 105-116.

Kaltz, O. & Shykoff, J. A. 1998. Local adaptation in host-parasite systems. *Heredity* **81**: 361-370.

Karatayev, A. Y., Boltovskoy, D., Padilla, D. K. & Burlakova, L. E. 2007. The invasive bivalves *Dreissena polymorpha* and *Limnoperna fortunei*: parallels, contrasts, potential spread and invasion impacts. *Journal of Shellfish Research* **26**: 205-213.

Karvonen, A., Jokela, J. & Laine, A. L. 2019. Importance of sequence and timing in parasite coinfections. *Trends in Parasitology* **35**: 109-118.

Kaufmann, J., Lenz, T. L., Kalbe, M., Milinski, M. & Eizaguirre, C. 2017. A field reciprocal transplant experiment reveals asymmetric costs of migration between lake and river ecotypes of three-spined sticklebacks (*Gasterosteus aculeatus*). *Journal of Evolutionary Biology* **30**: 938-950.

Keller, R. P., Ermgassen, P. & Aldridge, D. C. 2009. Vectors and timing of freshwater invasions in Great Britain. *Conservation Biology* **23**: 1526-1534.

Keller, R. P., Geist, J., Jeschke, J. M. & Kühn, I. 2011. Invasive species in Europe: ecology, status, and policy. *Environmental Sciences Europe* **23**: 23.

Kelly, C. E., Kennedy, C. R. & Brown, J. A. 2000. Physiological status of wild European eels (*Anguilla anguilla*) infected with the parasitic nematode, *Anguillicola crassus*. *Parasitology* **120**: 195-202.

Kennedy, C. R. 2001. Metapopulation and community dynamics of helminth parasites of eels *Anguilla anguilla* in the River Exe system. *Parasitology* **122**: 689-698.

Kennedy, C. R. & Fitch, D. J. 1990. Colonization, larval survival and epidemiology of the nematode *Anguillicola crassus*, parasitic in the eel, *Anguilla anguilla*, in Britain. *Journal of Fish Biology* **36**: 117-131.

Keppel, M., Dangel, K. C. & Sures, B. 2014. Comparison of infection success, development and swim bladder pathogenicity of two congeneric *Anguillicola* species in experimentally infected *Anguilla anguilla* and *A. japonica*. *Parasitology Research* **113**: 3727-3735.

Keppel, M., Dangel, K. C. & Sures, B. 2016. The Hsp70 response of *Anguillicola* species to host-specific stressors. *Parasitology Research* **115**: 2149-2154.

Kerr, P. J. 2012. Myxomatosis in Australia and Europe: a model for emerging infectious diseases. *Antiviral Research* **93**: 387-415.

Kirk, R. S. 2003. The impact of *Anguillicola crassus* on European eels. *Fisheries Management and Ecology* **10**: 385-394.

Kleindorfer, S. & Dudaniec, R. Y. 2016. Host-parasite ecology, behavior and genetics: a review of the introduced fly parasite *Philornis downsi* and its Darwin's finch hosts. *BMC Zoology* **1**: 1.

Klemme, I. & Karvonen, A. 2017. Vertebrate defense against parasites: interactions between avoidance, resistance, and tolerance. *Ecology and Evolution* **7**: 561-571.

Knopf, K. 2006. The swimbladder nematode *Anguillicola crassus* in the European eel *Anguilla anguilla* and the Japanese eel *Anguilla japonica*: differences in susceptibility and immunity between a recently colonized host and the original host. *Journal of Helminthology* **80**: 129-136.

Knopf, K. & Lucius, R. 2008. Vaccination of eels (*Anguilla japonica* and *Anguilla anguilla*) against *Anguillicola crassus* with irradiated L$_3$. *Parasitology* **135**: 633-640.

Knopf, K. & Mahnke, M. 2004. Differences in susceptibility of the European eel (*Anguilla anguilla*) and the Japanese eel (*Anguilla japonica*) to the swim-bladder nematode *Anguillicola crassus*. *Parasitology* **129**: 491-496.

Knopf, K., Naser, K., van der Heijden, M. H. T. & Taraschewski, H. 2000. Humoral immune response of European eel *Anguilla anguilla* experimentally infected with *Anguillicola crassus*. *Diseases of Aquatic Organisms* **42**: 61-69.

Knopf, K., Würtz, J., Sures, B. & Taraschewski, H. 1998. Impact of low water temperature on the development of *Anguillicola crassus* in the final host *Anguilla anguilla*. *Diseases of Aquatic Organisms* **33**: 143-149.

Knudsen, R., Amundsen, P. A. & Klemetsen, A. 2003. Inter- and intra-morph patterns in helminth communities of sympatric whitefish morphs. *Journal of Fish Biology* **62**: 847-859.

Kock, R. A., Woodford, M. H. & Rossiter, P. B. 2010. Disease risks associated with the translocation of wildlife. *Revue Scientifique Et Technique-Office International Des Epizooties* **29**: 329-350.

Kohler, S. L. & Wiley, M. J. 1997. Pathogen outbreaks reveal large-scale effects of competition in stream communities. *Ecology* **78**: 2164-2176.

Kopp, K. & Jokela, J. 2007. Resistant invaders can convey benefits to native species. *Oikos* **116**: 295-301.

Koskella, B. & Lively, C. M. 2009. Evidence for negative frequency-dependent selection during experimental coevolution of a freshwater snail and a sterilizing trematode. *Evolution* **63**: 2213-2221.

Koval, M. 2006. Claudins - Key pieces in the tight junction puzzle. *Cell Communication and Adhesion* **13**: 127-138.

Kowal, K., Silver, R., Slawinska, E., Bielecki, M., Chyczewski, L. & Kowal-Bielecka, O. 2011. CD163 and its role in inflammation. *Folia Histochemica Et Cytobiologica* **49**: 365-374.

Kozubikova, E., Petrusek, A., Duris, Z. & Oidtmann, B. 2007. *Aphanomyces astaci*, the crayfish plague pathogen, may be a common cause of crayfish mass mortalities in the Czech Republic. *Bulletin of the European Association of Fish Pathologists* **27**: 79-82.

Kuchta, R., Choudhury, A. & Scholz, T. 2018. Asian fish tapeworm: the most successful invasive parasite in freshwaters. *Trends in Parasitology* **34**: 511-523.

Kumar, G., Abd-Elfattah, A. & El-Matbouli, M. 2015. Identification of differentially expressed genes of brown trout (*Salmo trutta*) and rainbow trout (*Oncorhynchus mykiss*) in response to *Tetracapsuloides bryosalmonae* (Myxozoa). *Parasitology Research* **114**: 929-939.

Kurze, C., Mayack, C., Hirche, F., Stangl, G. I., Le Conte, Y., Kryger, P. & Moritz, R. F. A. 2016. *Nosema* spp. infections cause no energetic stress in tolerant honeybees. *Parasitology Research* **115**: 2381-2388.

Langmead, B. & Salzberg, S. L. 2012. Fast gapped-read alignment with Bowtie 2. *Nature Methods* **9**: 357-U54.

Langmead, B., Trapnell, C., Pop, M. & Salzberg, S. L. 2009. Ultrafast and memory-efficient alignment of short DNA sequences to the human genome. *Genome Biology* **10**: 10.

Layer, G., Reichelt, J., Jahn, D. & Heinz, D. W. 2010. Structure and function of enzymes in heme biosynthesis. *Protein Science* **19**: 1137-1161.

Le Conte, Y., Ellis, M. & Ritter, W. 2010. *Varroa* mites and honey bee health: can *Varroa* explain part of the colony losses? *Apidologie* **41**: 353-363.

Le Cren, E. D. 1951. The length-weight relationship and seasonal cycle in gonad weight and condition in the perch (*Perca fluviatilis*). *Journal of Animal Ecology* **20**: 201-219.

Leek, J. T., Scharpf, R. B., Bravo, H. C., Simcha, D., Langmead, B., Johnson, W. E., Geman, D., Baggerly, K. & Irizarry, R. A. 2010. Tackling the widespread and critical impact of batch effects in high-throughput data. *Nature Reviews Genetics* **11**: 733-739.

Lefebvre, F., Contournet, P. & Crivelli, A. J. 2007. Interaction between the severity of the infection by the nematode *Anguillicola crassus* and the tolerance to hypoxia in the European eel *Anguilla anguilla*. *Acta Parasitologica* **52**: 171-175.

Lefebvre, F., Contournet, P., Priour, F., Soulas, O. & Crivelli, A. J. 2002. Spatial and temporal variation in *Anguillicola crassus* counts: results of a 4 year survey of eels in Mediterranean lagoons. *Diseases of Aquatic Organisms* **50**: 181-188.

Lefebvre, F. & Crivelli, A. 2004. Anguillicolosis: dynamics of the infection over two decades. *Diseases of Aquatic Organisms* **62**: 227-232.

Lefebvre, F., Fazio, G., Mounaix, B. & Crivelli, A. J. 2013. Is the continental life of the European eel *Anguilla anguilla* affected by the parasitic invader *Anguillicoloides crassus*? *Proceedings of the Royal Society B-Biological Sciences* **280**.

Lefebvre, F., Schuster, T., Munderle, M., Hine, M. & Poulin, R. 2004. Anguillicolosis in the short-finned eel *Anguilla australis*: epidemiology and pathogenicity. *New Zealand Journal of Marine and Freshwater Research* **38**: 577-583.

Lefèvre, T., Lebarbenchon, C., Gauthier-Clerc, M., Missé, D., Poulin, R. & Thomas, F. 2009. The ecological significance of manipulative parasites. *Trends in Ecology & Evolution* **24**: 41-48.

Lenz, T. L., Eizaguirre, C., Rotter, B., Kalbe, M. & Milinski, M. 2013. Exploring local immunological adaptation of two stickleback ecotypes by experimental infection and transcriptome-wide digital gene expression analysis. *Molecular Ecology* **22**: 774-786.

Li, B. & Dewey, C. N. 2011. RSEM: accurate transcript quantification from RNA-Seq data with or without a reference genome. *BMC Bioinformatics* **12**: 16.

Li, B., Power, M. R. & Lin, T. J. 2006. De novo synthesis of early growth response factor-1 is required for the full responsiveness of mast cells to produce TNF and IL-13 by IgE and antigen stimulation. *Blood* **107**: 2814-2820.

Li, W. X., Arnott, S. A., Jones, K. M. M., Braicovich, P. E., de Buron, I., Wang, G. T. & Marcogliese, D. J. 2015. First record of paratenic hosts of the swimbladder nematode *Anguillicola crassus* in North America. *Journal of Parasitology* **101**: 529-535.

Limnander, A., Depeille, P., Freedman, T. S., Liou, J., Leitges, M., Kurosaki, T., Roose, J. P. & Weiss, A. 2011. STIM1, PKC-delta and RasGRP set a threshold for proapoptotic Erk signaling during B cell development. *Nature Immunology* **12**: 425-U81.

Lintermans, M. 2004. Human-assisted dispersal of alien freshwater fish in Australia. *New Zealand Journal of Marine and Freshwater Research* **38**: 481-501.

Lips, K. R., Brem, F., Brenes, R., Reeve, J. D., Alford, R. A., Voyles, J., Carey, C., Livo, L., Pessier, A. P. & Collins, J. P. 2006. Emerging infectious disease and the loss of biodiversity in a Neotropical amphibian community. *Proceedings of the National Academy of Sciences of the United States of America* **103**: 3165-3170.

Lochmiller, R. L. & Deerenberg, C. 2000. Trade-offs in evolutionary immunology: just what is the cost of immunity? *Oikos* **88**: 87-98.

Logan, C. A. & Somero, G. N. 2010. Transcriptional responses to thermal acclimation in the eurythermal fish *Gillichthys mirabilis* (Cooper 1864). *American Journal of Physiology-Regulatory Integrative and Comparative Physiology* **299**: R843-R852.

Lohman, B. K., Steinel, N. C., Weber, J. N. & Bolnick, D. I. 2017. Gene expression contributes to the recent evolution of host resistance in a model host parasite system. *Frontiers in Immunology* **8**: 12.

Lohoff, M., Giaisi, M., Kohler, R., Casper, B., Krammer, P. H. & Li-Weber, M. 2010. Early growth response protein-1 (Egr-1) is preferentially expressed in T helper type 2 (Th2) cells and is involved in acute transcription of the Th2 cytokine interleukin-4. *Journal of Biological Chemistry* **285**: 1643-1652.

Loiseau, C., Zoorob, R., Robert, A., Chastel, O., Julliard, R. & Sorci, G. 2011. Plasmodium relictum infection and MHC diversity in the house sparrow (*Passer domesticus*). *Proceedings of the Royal Society B-Biological Sciences* **278**: 1264-1272.

Love, M. I., Huber, W. & Anders, S. 2014. Moderated estimation of fold change and dispersion for RNA-seq data with DESeq2. *Genome Biology* **15**: 38.

Lowry, E., Rollinson, E. J., Laybourn, A. J., Scott, T. E., Aiello-Lammens, M. E., Gray, S. M., Mickley, J. & Gurevitch, J. 2013. Biological invasions: a field synopsis, systematic review, and database of the literature. *Ecology and Evolution* **3**: 182-196.

Lymbery, A. J., Morine, M., Kanani, H. G., Beatty, S. J. & Morgan, D. L. 2014. Co-invaders: the effects of alien parasites on native hosts. *International Journal for Parasitology: Parasites and Wildlife* **3**: 171-177.

MacColl, A. D. C. & Chapman, S. M. 2010. Parasites can cause selection against migrants following dispersal between environments. *Functional Ecology* **24**: 847-856.

Madsen, S. S., Engelund, M. B. & Cutler, C. P. 2015. Water transport and functional dynamics of aquaporins in osmoregulatory organs of fishes. *Biological Bulletin* **229**: 70-92.

Marzal, A., García-Longoria, L., Callirgos, J. M. C. & Sehgal, R. N. M. 2015. Invasive avian malaria as an emerging parasitic disease in native birds of Peru. *Biological Invasions* **17**: 39-45.

Mastitsky, S. E., Karatayev, A. Y., Burlakova, L. E. & Molloy, D. P. 2010. Parasites of exotic species in invaded areas: does lower diversity mean lower epizootic impact? *Diversity and Distributions* **16**: 798-803.

McKenzie, V. J. & Peterson, A. C. 2012. Pathogen pollution and the emergence of a deadly amphibian pathogen. *Molecular Ecology* **21**: 5151-5154.

McKnight, D. T., Schwarzkopf, L., Alford, R. A., Bower, D. S. & Zenger, K. R. 2017. Effects of emerging infectious diseases on host population genetics: a review. *Conservation Genetics* **18**: 1235-1245.

McMahon, S. B. & Monroe, J. G. 1996. The role of early growth response gene 1 (egr-1) in regulation of the immune response. *Journal of Leukocyte Biology* **60**: 159-166.

Mecklenbräuker, I., Saijo, K., Zheng, N. Y., Leitges, M. & Tarakhovsky, A. 2002. Protein kinase C delta controls self-antigen-induced B-cell tolerance. *Nature* **416**: 860-865.

Medzhitov, R., Schneider, D. S. & Soares, M. P. 2012. Disease tolerance as a defense strategy. *Science* **335**: 936-941.

Meteyer, C. U., Barber, D. & Mandl, J. N. 2012. Pathology in euthermic bats with white nose syndrome suggests a natural manifestation of immune reconstitution inflammatory syndrome. *Virulence* **3**: 583-588.

Miller, M. J., Feunteun, E. & Tsukamoto, K. 2016. Did a "perfect storm" of oceanic changes and continental anthropogenic impacts cause northern hemisphere anguillid recruitment reductions? *Ices Journal of Marine Science* **73**: 43-56.

Miller, M. P. & Vincent, E. R. 2008. Rapid natural selection for resistance to an introduced parasite of rainbow trout. *Evolutionary Applications* **1**: 336-341.

Miyamoto, A., Nakayama, K., Imaki, H., Hirose, S., Jiang, Y., Abe, M., Tsukiyama, T., Nagahama, H., Ohno, S., Hatakeyama, S. & Nakayama, K. I. 2002. Increased proliferation of B cells and auto-immunity in mice lacking protein kinase C delta. *Nature* **416**: 865-869.

Mollot, G., Pantel, J. H. & Romanuk, T. N. (2017) The effects of invasive species on the decline in species richness: a global meta-analysis. In: *Networks of invasion: a synthesis of concepts*, Vol. 56 (Bohan, D. A., Dumbrell, A. J. & Massol, F., eds.). pp. 61-84 Advances in Ecological Research. Elsevier Academic Press Inc, San Diego.

Molnár, K. 1993. Effect of decreased oxygen-content on eels (*Anguilla anguilla*) infected by *Anguillicola crassus* (Nematoda, Dracunculoidea). *Acta Veterinaria Hungarica* **41**: 349-360.

Molnár, K. 1994. Formation of parasitic nodules in the swimbladder and intestinal walls of the eel *Anguilla anguilla* due to infections with larval stages of *Anguillicola crassus*. *Diseases of Aquatic Organisms* **20**: 163-170.

Molnár, K., Baska, F., Csaba, G., Glávits, R. & Székely, C. 1993. Pathological and histopathological studies of the swimbladder of eels *Anguilla anguilla* infected by *Anguillicola crassus* (Nematoda: Dracunculoidea). *Diseases of Aquatic Organisms* **15**: 41-50.

Molnár, K., Székely, C. & Baska, F. 1991. Mass mortality of eel in Lake Balaton due to *Anguillicola crassus* infection. *Bulletin of the European Association for Fish Pathology* **11**: 211-212.

Moravec, F., Dicave, D., Orecchia, P. & Paggi, L. 1994a. Experimental observations on the development of *Anguillicola crassus* (Nematoda: Dracunculoidea) in its definitive host, *Anguilla anguilla* (Pisces). *Folia Parasitologica* **41**: 138-148.

Moravec, F., Dicave, D., Orecchia, P. & Paggi, L. 1994b. Present occurrence of *Anguillicola novaezelandiae* (Nematoda, Dracunculoidea) in Europe and its development in the intermediate host. *Folia Parasitologica* **41**: 203-208.

Moravec, F. & Konecny, R. 1994. Some new data on the intermediate and paratenic hosts of the nematode *Anguillicola crassus* Kuwahara, Niimi et Itagaki, 1974 (Dracunculoidea), a swimbladder parasite of eels. *Folia Parasitologica* **41**: 65-70.

Moreau, E. & Chauvin, A. 2010. Immunity against helminths: interactions with the host and the intercurrent infections. *Journal of Biomedicine and Biotechnology*: 9.

Mori, E., Sala, J. P., Fattorini, N., Menchetti, M., Montalvo, T. & Senar, J. C. 2019. Ectoparasite sharing among native and invasive birds in a metropolitan area. *Parasitology Research* **118**: 399-409.

Morris, A., Hewitt, C. & Young, S. 1994. The major histocompatibility complex: its genes and their roles in antigen presentation. *Molecular Aspects of Medicine* **15**: 377-503.

Mulcahy, G., O'Neill, S., Fanning, J., McCarthy, E. & Sekiya, M. 2005. Tissue migration by parasitic helminths - an immunoevasive strategy? *Trends in Parasitology* **21**: 273-277.

Murphy, K. 2012. *Janeway's Immunobiology,* 8th ed. Garland Science, Taylor & Fancis Group, LLC, New York, USA.

Murphy, K. & Weaver, C. 2017. *Janeway's immunobiology,* 9th ed. Garland Science, Taylor & Francis Group, LLC, New York, USA.

Möller, H., Holst, S., Lüchtenberg, H. & Petersen, F. 1991. Infection of eel *Anguilla anguilla* from the River Elbe estuary with 2 nematodes, *Anguillicola crassus* and *Pseudoterranova decipiens.* *Diseases of Aquatic Organisms* **11**: 193-199.

Münderle, M., Taraschewski, H., Klar, B., Chang, C. W., Shiao, J. C., Shen, K. N., He, J. T., Lin, S. H. & Tzeng, W. N. 2006. Occurrence of *Anguillicola crassus* (Nematoda : Dracunculoidea) in Japanese eels *Anguilla japonica* from a river and an aquaculture unit in SW Taiwan. *Diseases of Aquatic Organisms* **71**: 101-108.

Nagasawa, K., Kim, Y. G. & Hirose, H. 1994. *Anguillicola crassus* and *A. globiceps* (Nematoda: Dracunculoidea) parasitic in the swimbladder of eels (*Anguilla japonica* and *A. anguilla*) in East Asia - a review. *Folia Parasitologica* **41**: 127-137.

Nelson, F. B. L., Brown, G. P., Shilton, C. & Shine, R. 2015. Host-parasite interactions during a biological invasion: The fate of lungworms (*Rhabdias* spp.) inside native and novel anuran hosts. *International Journal for Parasitology-Parasites and Wildlife* **4**: 206-215.

Neto, A. F., Costa, J. L., Costa, M. J. & Domingos, I. 2010. Epidemiology and pathology of *Anguillicoloides crassus* in European eel *Anguilla anguilla* from the Tagus estuary (Portugal). *Diseases of Aquatic Organisms* **88**: 225-233.

Newbold, L. R., Hockley, F. A., Williams, C. F., Cable, J., Reading, A. J., Auchterlonie, N. & Kemp, P. S. 2015. Relationship between European eel *Anguilla anguilla* infection with non-native parasites and swimming behaviour on encountering accelerating flow. *Journal of Fish Biology* **86**: 1519-1533.

Nielsen, M. E. 1999. An enhanced humoral immune response against the swimbladder nematode, *Anguillicola crassus*, in the Japanese eel, *Anguilla japonica*, compared with the European eel, *A. anguilla*. *Journal of Helminthology* **73**: 227-232.

Nikinmaa, M. & Rytkönen, K. T. 2012. From genomes to functions in aquatic biology. *Marine Genomics* **5**: 1-6.

Nimeth, K., Zwerger, P., Wurtz, J., Salvenmoser, W. & Pelster, B. 2000. Infection of the glass-eel swimbladder with the nematode *Anguillicola crassus*. *Parasitology* **121**: 75-83.

Norton, J., Rollinson, D. & Lewis, J. W. 2005. Epidemiology of *Anguillicola crassus* in the European eel (*Anguilla anguilla*) from two rivers in southern England. *Parasitology* **130**: 679-686.

Okamura, B. & Feist, S. W. 2011. Emerging diseases in freshwater systems. *Freshwater Biology* **56**: 627-637.

Oksanen, J., Blanchet, F. G., Friendly, M., Kindt, R., Legendre, P., McGlinn, D., Minchin, P. R., O'Hara, R. B., Simpson, G. L., Solymos, P., Stevens, M. H. H., Szoecs, E. & Wagner, H. (2017) Vegan: community ecology package. pp.

Padilla, D. K. & Williams, S. L. 2004. Beyond ballast water: aquarium and ornamental trades as sources of invasive species in aquatic ecosystems. *Frontiers in Ecology and the Environment* **2**: 131-138.

Page, K. R., Scott, A. L. & Manabe, Y. C. 2006. The expanding realm of heterologous immunity: friend or foe? *Cellular Microbiology* **8**: 185-196.

Palstra, A. P., Heppener, D. F. M., van Ginneken, V. J. T., Székely, C. & van den Thillart, G. 2007. Swimming performance of silver eels is severely impaired by the swim-bladder parasite *Anguillicola crassus*. *Journal of Experimental Marine Biology and Ecology* **352**: 244-256.

Palstra, A. P. & van den Thillart, G. 2010. Swimming physiology of European silver eels (*Anguilla anguilla* L.): energetic costs and effects on sexual maturation and reproduction. *Fish Physiology and Biochemistry* **36**: 297-322.

Pan, Y. B., Wan, J. X., Liu, Y. P., Yang, Q., Liang, W., Singhal, P. C., Saleem, M. A. & Ding, G. H. 2014. sPLA2 IB induces human podocyte apoptosis via the M-type phospholipase A2 receptor. *Scientific Reports* **4**: 11.

Paoli, M., Marles-Wright, J. & Smith, A. 2002. Structure-function relationships in heme-proteins. *DNA and Cell Biology* **21**: 271-280.

Park, D. W., Kim, J. R., Kim, S. Y., Sonn, J. K., Bang, O. S., Kang, S. S., Kim, J. H. & Baek, S. H. 2003. Akt as a mediator of secretory phospholipase A_2 receptor-involved inducible nitric oxide synthase expression. *Journal of Immunology* **170**: 2093-2099.

Parkin, J. & Cohen, B. 2001. An overview of the immune system. *Lancet* **357**: 1777-1789.

Pavey, S. A., Bernatchez, L., Aubin-Horth, N. & Landry, C. R. 2012. What is needed for next-generation ecological and evolutionary genomics? *Trends in Ecology & Evolution* **27**: 673-678.

Pedersen, A. B. & Fenton, A. 2007. Emphasizing the ecology in parasite community ecology. *Trends in Ecology & Evolution* **22**: 133-139.

Peeler, E. J. & Feist, S. W. 2011. Human intervention in freshwater ecosystems drives disease emergence. *Freshwater Biology* **56**: 705-716.

Peeler, E. J., Oidtmann, B. C., Midtlyng, P. J., Miossec, L. & Gozlan, R. E. 2011. Non-native aquatic animals introductions have driven disease emergence in Europe. *Biological Invasions* **13**: 1291-1303.

Pegg, J., Andreou, D., Williams, C. F. & Britton, J. R. 2015. Head morphology and piscivory of European eels, *Anguilla anguilla*, predict their probability of infection by the invasive parasitic nematode *Anguillicoloides crassus*. *Freshwater Biology* **60**: 1977-1987.

Pelster, B. 2015. Swimbladder function and the spawning migration of the European eel *Anguilla anguilla*. *Frontiers in Physiology* **5**: 10.

Pelster, B., Schneebauer, G. & Dirks, R. P. 2016. *Anguillicola crassus* infection significantly affects the silvering related modifications in steady state mRNA levels in gas gland tissue of the European eel. *Frontiers in Physiology* **7**: 13.

Penczykowski, R. M., Forde, S. E. & Duffy, M. A. 2011. Rapid evolution as a possible constraint on emerging infectious diseases. *Freshwater Biology* **56**: 689-704.

Peters, G. & Hartmann, F. 1986. *Anguillicola*, a parasitic nematode of the swim bladder spreading among eel populations in Europe. *Diseases of Aquatic Organisms* **1**: 229-230.

Pfaffl, M. W. 2001. A new mathematical model for relative quantification in real-time RT-PCR. *Nucleic Acids Research* **29**: 6.

Pizzatto, L., Kelehear, C., Dubey, S., Barton, D. & Shine, R. 2012. Host-parasite relationship during a biologic invasion: 75 years postinvasion, cane toads and sympatric Australian frogs retain separate lungworm faunas. *Journal of Wildlife Diseases* **48**: 951-961.

Plotkin, J. B. 2010. Transcriptional regulation is only half the story. *Molecular Systems Biology* **6**: 2.

Polo-Cavia, N., Lopez, P. & Martin, J. 2010. Competitive interactions during basking between native and invasive freshwater turtle species. *Biological Invasions* **12**: 2141-2152.

Poorten, T. J. & Rosenblum, E. B. 2016. Comparative study of host response to chytridiomycosis in a susceptible and a resistant toad species. *Molecular Ecology* **25**: 5663-5679.

Poulin, R. 1999. The functional importance of parasites in animal communities: many roles at many levels? *International Journal for Parasitology* **29**: 903-914.

Press, C. M. & Evensen, O. 1999. The morphology of the immune system in teleost fishes. *Fish & Shellfish Immunology* **9**: 309-318.

Preston, D. L., Mischler, J. A., Townsend, A. R. & Johnson, P. T. J. 2016. Disease ecology meets ecosystem science. *Ecosystems* **19**: 737-748.

Primmer, C. R., Papakostas, S., Leder, E. H., Davis, M. J. & Ragan, M. A. 2013. Annotated genes and nonannotated genomes: Cross-species use of Gene Ontology in ecology and evolution research. *Molecular Ecology* **22**: 3216-3241.

Pujolar, J. M., Jacobsen, M. W., Als, T. D., Frydenberg, J., Munch, K., Jonsson, B., Jian, J. B., Cheng, L., Maes, G. E., Bernatchez, L. & Hansen, M. M. 2014. Genome-wide single-generation signatures of local selection in the panmictic European eel. *Molecular Ecology* **23**: 2514-2528.

Quezada, S. A., Jarvinen, L. Z., Lind, E. E. & Noelle, R. J. 2004. CD40/CD154 interactions at the interface of tolerance and immunity. *Annual Review of Immunology* **22**: 307-328.

R Core Team (2016) R: A language and environment for statistical computing. pp. R Foundation for Statistical Computing, Vienna, Austria.

R Core Team (2019) R: A language and environment for statistical computing. pp. R Foundation for Statistical Computing, Vienna, Austria.

Raberg, L., Graham, A. L. & Read, A. F. 2009. Decomposing health: tolerance and resistance to parasites in animals. *Philosophical Transactions of the Royal Society B-Biological Sciences* **364**: 37-49.

Raberg, L., Sim, D. & Read, A. F. 2007. Disentangling genetic variation for resistance and tolerance to infectious diseases in animals. *Science* **318**: 812-814.

Rauque, C., Viozzi, G., Flores, V., Vega, R., Waicheim, A. & Salgado-Maldonado, G. 2018. Helminth parasites of alien freshwater fishes in Patagonia (Argentina). *International Journal for Parasitology-Parasites and Wildlife* **7**: 369-379.

Read, A. F., Graham, A. L. & Raberg, L. 2008. Animal defenses against infectious agents: is damage control more important than pathogen control? *Plos Biology* **6**: 2638-2641.

Reche, P. A. & Reinherz, E. L. 2003. Sequence variability analysis of human class I and class II MHC molecules: Functional and structural correlates of amino acid polymorphisms. *Journal of Molecular Biology* **331**: 623-641.

Reid, B. L., Hernández, K. L., Frangópulos, M., Bauer, G., Lorca, M., Kilroy, C. & Spaulding, S. 2012. The invasion of the freshwater diatom *Didymosphenia geminata* in Patagonia: prospects, strategies, and implications for biosecurity of invasive microorganisms in continental waters. *Conservation Letters* **5**: 432-440.

Robledo, D., Ronza, P., Harrison, P. W., Losada, A. P., Bermudez, R., Pardo, B. G., Redondo, M. J., Sitja-Bobadilla, A., Quiroga, M. I. & Martinez, P. 2014. RNA-seq analysis reveals significant transcriptome changes in turbot (*Scophthalmus maximus*) suffering severe enteromyxosis. *BMC Genomics* **15**: 17.

Roeder, P., Mariner, J. & Kock, R. 2013. Rinderpest: the veterinary perspective on eradication. *Philosophical Transactions of the Royal Society B-Biological Sciences* **368**: 12.

Ronza, P., Robledo, D., Bermudez, R., Losada, A. P., Pardo, B. G., Sitja-Bobadilla, A., Quiroga, M. I. & Martinez, P. 2016. RNA-seq analysis of early enteromyxosis in turbot (*Scophthalmus maximus*): new insights into parasite invasion and immune evasion strategies. *International Journal for Parasitology* **46**: 507-517.

Rosenblum, E. B., Poorten, T. J., Settles, M. & Murdoch, G. K. 2012. Only skin deep: shared genetic response to the deadly chytrid fungus in susceptible frog species. *Molecular Ecology* **21**: 3110-3120.

Sadd, B. M. & Schmid-Hempel, P. 2009. Principles of ecological immunology. *Evolutionary Applications* **2**: 113-121.

Samuel, M. D., Woodworth, B. L., Atkinson, C. T., Hart, P. J. & LaPointe, D. A. 2015. Avian malaria in Hawaiian forest birds: infection and population impacts across species and elevations. *Ecosphere* **6**: 21.

Sarabian, C., Curtis, V. & McMullan, R. 2018. Evolution of pathogen and parasite avoidance behaviours. *Philosophical Transactions of the Royal Society B-Biological Sciences* **373**: 7.

Schabuss, M., Kennedy, C. R., Konecny, R., Grillitsch, B., Reckendorfer, W., Schiemer, F. & Herzig, A. 2005. Dynamics and predicted decline of *Anguillicola crassus* infection in European eels, *Anguilla anguilla*, in Neusiedler See, Austria. *Journal of Helminthology* **79**: 159-167.

Scharsack, J. P. & Kalbe, M. 2014. Differences in susceptibility and immune responses of three-spined sticklebacks (*Gasterosteus aculeatus*) from lake and river ecotypes to sequential infections with the eye fluke *Diplostomum pseudospathaceum*. *Parasites & Vectors* **7**: 10.

Scheele, B., Pasmans, F., Skerratt, L. F., Berger, L., Martel, A., Beukema, W., Acevedo, A. A., Burrowes, P. A., Carvalho, T., Catenazzi, A., De la Riva, I., Fisher, M. C., Flechas, S. V., Foster, C. N., Frías-Álvarez, P., Garner, T. W. J., Gratwicke, B., Guayasamin, J. M., Hirschfeld, M., Kolby, J. E., Kosch, T. A., La Marca, E., Lindenmayer, D. B., Lips, K. R., Longo, A. V., Maneyro, R., McDonald, C. A., Mendelson, J., Palacios-Rodriguez, P., Parra-Olea, G., Richards-Zawacki, C. L., Rödel, M. O., Rovito, S. M., Soto-Azat, C., Toledo, L. F., Voyles, J., Weldon, C., Whitfield, S. M., Wilkinson, M., Zamudio, K. R. & Canessa, S. 2019. Amphibian fungal panzootic causes catastrophic and ongoing loss of biodiversity. *Science* **363**: 1459-+.

Schmid-Hempel, P. 2003. Variation in immune defence as a question of evolutionary ecology. *Proceedings of the Royal Society B-Biological Sciences* **270**: 357-366.

Schmidt, J. 1923. The breeding places of the eel. *Philosophical Transactions of the Royal Society of London. Series B, Containing Papers of a Biological Character* **211**: 179-208.

Schneebauer, G., Dirks, R. P. & Pelster, B. 2017. *Anguillicola crassus* infection affects mRNA expression levels in gas gland tissue of European yellow and silver eel. *Plos One* **12**: 26.

Schneider, D. S. & Ayres, J. S. 2008. Two ways to survive infection: what resistance and tolerance can teach us about treating infectious diseases. *Nature Reviews Immunology* **8**: 889-895.

Schoebel, C. N., Wolinska, J. & Spaak, P. 2010. Higher parasite resistance in *Daphnia* populations with recent epidemics. *Journal of Evolutionary Biology* **23**: 2370-2376.

Schoener, E. R., Banda, M., Howe, L., Castro, I. C. & Alley, M. R. 2014. Avian malaria in New Zealand. *New Zealand Veterinary Journal* **62**: 189-198.

Schulenburg, H., Kurtz, J., Moret, Y. & Siva-Jothy, M. T. 2009. Ecological immunology. *Philosophical Transactions of the Royal Society B-Biological Sciences* **364**: 3-14.

Sheath, D. J., Williams, C. F., Reading, A. J. & Britton, J. R. 2015. Parasites of non-native freshwater fishes introduced into England and Wales suggest enemy release and parasite acquisition. *Biological Invasions* **17**: 2235-2246.

Sheldon, B. C. & Verhulst, S. 1996. Ecological immunology: costly parasite defences and trade-offs in evolutionary ecology. *Trends in Ecology & Evolution* **11**: 317-321.

Shirakashi, S., Teruya, K. & Ogawa, K. 2008. Altered behaviour and reduced survival of juvenile olive flounder, *Paralichthys olivaceus*, infected by an invasive monogenean, *Neoheterobothrium hirame*. *International Journal for Parasitology* **38**: 1513-1522.

Simberloff, D., Martin, J. L., Genovesi, P., Maris, V., Wardle, D. A., Aronson, J., Courchamp, F., Galil, B., García-Berthou, E., Pascal, M., Pysek, P., Sousa, R., Tabacchi, E. & Vila, M. 2013. Impacts of biological invasions: what's what and the way forward. *Trends in Ecology & Evolution* **28**: 58-66.

Sjöberg, N. B., Petersson, E., Wickström, H. & Hansson, S. 2009. Effects of the swimbladder parasite *Anguillicola crassus* on the migration of European silver eels *Anguilla anguilla* in the Baltic Sea. *Journal of Fish Biology* **74**: 2158-2170.

Skerratt, L. F., Berger, L., Speare, R., Cashins, S., McDonald, K. R., Phillott, A. D., Hines, H. B. & Kenyon, N. 2007. Spread of chytridiomycosis has caused the rapid global decline and extinction of frogs. *Ecohealth* **4**: 125-134.

Skugor, S., Glover, K. A., Nilsen, F. & Krasnov, A. 2008. Local and systemic gene expression responses of Atlantic salmon (*Salmo salar* L.) to infection with the salmon louse (*Lepeophtheirus salmonis*). *BMC Genomics* **9**: 18.

Smith, K. F., Acevedo-Whitehouse, K. & Pedersen, A. B. 2009. The role of infectious diseases in biological conservation. *Animal Conservation* **12**: 1-12.

Sorci, G. 2013. Immunity, resistance and tolerance in bird-parasite interactions. *Parasite Immunology* **35**: 350-361.

Staley, M. & Bonneaud, C. 2015. Immune responses of wild birds to emerging infectious diseases. *Parasite Immunology* **37**: 242-254.

Stanford, B. C. M. & Rogers, S. M. 2018. R(NA)-tistic expression: the art of matching unknown mRNA and proteins to environmental response in ecological genomics. *Molecular Ecology* **27**: 827-830.

Stutz, W. E., Schmerer, M., Coates, J. L. & Bolnick, D. I. 2015. Among-lake reciprocal transplants induce convergent expression of immune genes in threespine stickleback. *Molecular Ecology* **24**: 4629-4646.

Sures, B. & Knopf, K. 2004. Parasites as a threat to freshwater eels? *Science* **304**: 208-209.

Sures, B. & Streit, B. 2001. Eel parasite diversity and intermediate host abundance in the River Rhine, Germany. *Parasitology* **123**: 185-191.

Svoboda, J., Mrugala, A., Kozubíková-Balcarová, E. & Petrusek, A. 2017. Hosts and transmission of the crayfish plague pathogen *Aphanomyces astaci*: a review. *Journal of Fish Diseases* **40**: 127-140.

Székely, C. 1994. Paratenic hosts for the parasitic nematode *Anguillicola crassus* in Lake Balaton, Hungary. *Diseases of Aquatic Organisms* **18**: 11-20.

Taraschewski, H. 2006. Hosts and parasites as aliens. *Journal of Helminthology* **80**: 99-128.

Telfer, S., Bown, K. J., Sekules, R., Begon, I., Hayden, T. & Birtles, R. 2005. Disruption of a host-parasite system following the introduction of an exotic host species. *Parasitology* **130**: 661-668.

Tesch, F.-W. 2003. *The eel,* 3rd ed. Blackwell Science Ltd, Oxford, UK.

Thomas, F., Mete, K., Helluy, S., Santalla, F., Verneau, O., DeMeeüs, T., Cézilly, F. & Renaud, F. 1997. Hitch-hiker parasites or how to benefit from the strategy of another parasite. *Evolution* **51**: 1316-1318.

Thomas, K. & Ollevier, F. 1992. Paratenic hosts of the swimbladder nematode *Anguillicola crassus*. *Diseases of Aquatic Organisms* **13**: 165-174.

Todd, E. V., Black, M. A. & Gemmell, N. J. 2016. The power and promise of RNA-seq in ecology and evolution. *Molecular Ecology* **25**: 1224-1241.

Tompkins, D. M., Carver, S., Jones, M. E., Krkosek, M. & Skerratt, L. F. 2015. Emerging infectious diseases of wildlife: a critical perspective. *Trends in Parasitology* **31**: 149-159.

Uribe, C., Folch, H., Enriquez, R. & Moran, G. 2011. Innate and adaptive immunity in teleost fish: a review. *Veterinarni Medicina* **56**: 486-503.

van Banning, P. & Haenen, O. L. M. 1990. Effects of the swimbladder nematode *Anguillicola crassus* in wild and farmed eel, *Anguilla anguilla*. *Pathology in Marine Science*: 317-330.

van den Thillart, G., van Ginneken, V., Korner, F., Heijmans, R., Van der Linden, R. & Gluvers, A. 2004. Endurance swimming of European eel. *Journal of Fish Biology* **65**: 312-318.

Van Gorp, H., Delputte, P. L. & Nauwynck, H. J. 2010. Scavenger receptor CD163, a Jack-of-all-trades and potential target for cell-directed therapy. *Molecular Immunology* **47**: 1650-1660.

van Riper, C., van Riper, S. G., Goff, M. L. & Laird, M. 1986. The epizootiology and ecological significance of malaria in Hawaiian land birds. *Ecological Monographs* **56**: 327-344.

Viney, M. E., Riley, E. M. & Buchanan, K. L. 2005. Optimal immune responses: immunocompetence revisited. *Trends in Ecology & Evolution* **20**: 665-669.

Vralstad, T., Strand, D. A., Grandjean, F., Kvellestad, A., Hastein, T., Knutsen, A. K., Taugbol, T. & Skaar, I. 2014. Molecular detection and genotyping of *Aphanomyces astaci* directly from preserved crayfish samples uncovers the Norwegian crayfish plague disease history. *Veterinary Microbiology* **173**: 66-75.

Weber, J. N., Kalbe, M., Shim, K. C., Erin, N. I., Steinel, N. C., Ma, L. & Bolnick, D. I. 2017. Resist globally, infect locally: a transcontinental test of adaptation by stickleback and their tapeworm parasite. *American Naturalist* **189**: 43-57.

Weclawski, U., Heitlinger, E. G., Baust, T., Klar, B., Petney, T., Han, Y. S. & Taraschewski, H. 2013. Evolutionary divergence of the swim bladder nematode *Anguillicola crassus* after colonization of a novel host, *Anguilla anguilla*. *BMC Evolutionary Biology* **13**.

Weclawski, U., Heitlinger, E. G., Baust, T., Klar, B., Petney, T., Han, Y. S. & Taraschewski, H. 2014. Rapid evolution of *Anguillicola crassus* in Europe: species diagnostic traits are plastic and evolutionarily labile. *Frontiers in Zoology* **11**: 9.

Wegner, K. M., Kalbe, M., Rauch, G., Kurtz, J., Schaschl, H. & Reusch, T. B. H. 2006. Genetic variation in MHC class II expression and interactions with MHC sequence polymorphism in three-spined sticklebacks. *Molecular Ecology* **15**: 1153-1164.

Wegner, K. M., Kalbe, M. & Reusch, T. B. H. 2007. Innate versus adaptive immunity in sticklebacks: evidence for trade-offs from a selection experiment. *Evolutionary Ecology* **21**: 473-483.

Whyte, S. K. 2007. The innate immune response of finfish - A review of current knowledge. *Fish & Shellfish Immunology* **23**: 1127-1151.

Wielgoss, S., Taraschewski, H., Meyer, A. & Wirth, T. 2008. Population structure of the parasitic nematode *Anguillicola crassus*, an invader of declining North Atlantic eel stocks. *Molecular Ecology* **17**: 3478-3495.

Windsor, D. A. 1998. Most of the species on Earth are parasites. *International Journal for Parasitology* **28**: 1939-1941.

Wolinska, J. & Spaak, P. 2009. The cost of being common: evidence from natural *Daphnia* populations. *Evolution* **63**: 1893-1901.

Wood, C. L., Byers, J. E., Cottingham, K. L., Altman, I., Donahue, M. J. & Blakeslee, A. M. H. 2007. Parasites alter community structure. *Proceedings of the National Academy of Sciences of the United States of America* **104**: 9335-9339.

Woolhouse, M. E. J., Webster, J. P., Domingo, E., Charlesworth, B. & Levin, B. R. 2002. Biological and biomedical implications of the co-evolution of pathogens and their hosts. *Nature Genetics* **32**: 569-577.

Wysujack, K., Dorow, M. & Ubl, C. 2014. The infection of the European eel with the parasitic nematode *Anguillicoloides crassus* in inland and coastal waters of northern Germany. *Journal of Coastal Conservation* **18**: 121-130.

Würtz, J., Knopf, K. & Taraschewski, H. 1998. Distribution and prevalence of *Anguillicola crassus* (Nematoda) in eels *Anguilla anguilla* of the rivers Rhine and Naab, Germany. *Diseases of Aquatic Organisms* **32**: 137-143.

Würtz, J. & Taraschewski, H. 2000. Histopathological changes in the swimbladder wall of the European eel *Anguilla anguilla* due to infections with *Anguillicola crassus*. *Diseases of Aquatic Organisms* **39**: 121-134.

Würtz, J., Taraschewski, H. & Pelster, B. 1996. Changes in gas composition in the swimbladder of the European eel (*Anguilla anguilla*) infected with *Anguillicola crassus* (Nematoda). *Parasitology* **112**: 233-238.

Yu, A. S. L. 2015. Claudins and the Kidney. *Journal of the American Society of Nephrology* **26**: 11-19.

Zambrano, L., Scheffer, M. & Martínez-Ramos, M. 2001. Catastrophic response of lakes to benthivorous fish introduction. *Oikos* **94**: 344-350.

Zavodna, M., Sandland, G. J. & Minchella, D. J. 2008. Effects of intermediate host genetic background on parasite transmission dynamics: a case study using *Schistosoma mansoni*. *Experimental Parasitology* **120**: 57-61.

Zelmer, D. A. 1998. An evolutionary definition of parasitism. *International Journal for Parasitology* **28**: 531-533.

Zhang, F. K., Zhang, X. X., Elsheikha, H. M., He, J. J., Sheng, Z. A., Zheng, W. B., Ma, J. G., Huang, W. Y., Guo, A. J. & Zhu, X. Q. 2017. Transcriptomic responses of water buffalo liver to infection with the digenetic fluke *Fasciola gigantica*. *Parasites & Vectors* **10**: 13.

Zhang, Y., Liu, X. J., Zhang, W. Q. & Han, R. C. 2010. Differential gene expression of the honey bees *Apis mellifera* and *A. cerana* induced by *Varroa destructor* infection. *Journal of Insect Physiology* **56**: 1207-1218.

Chapter I

Table SI.1 List of all differentially expressed genes in the spleen and the head kidney. Annotations, where available, were obtained by blastx searches against UniProt and RefSeq databases. Genes that were differentially expressed in both tissues are shaded in gray. Log2FC is the $\log_2$ fold change between expression in infected and control individuals, Wald stat is the Wald statistic calculated by DESeq2, and Direction indicates up-regulation ($\uparrow$) or down-regulation ($\downarrow$) in infected individuals.

Gene ID	Log2FC	Wald stat	p-value	Adj. p-value	Direction	UniProt annotations	RefSeq annotation
Spleen							
TRINITY_DN164373_c3_g4	2.88	6.90	5.23E-12	1.01E-07	$\uparrow$		
TRINITY_DN163744_c5_g3	2.77	6.57	4.87E-11	5.64E-07	$\uparrow$	Retrovirus-related Pol polyprotein from type-1 retrotransposable element R2	PREDICTED: uncharacterized protein LOC580961 [Strongylocentrotus purpuratus]
TRINITY_DN168651_c5_g8	2.14	5.97	2.35E-09	2.26E-05	$\uparrow$	High affinity immunoglobulin gamma Fc receptor I	
TRINITY_DN164847_c1_g5	2.37	5.56	2.71E-08	1.96E-04	$\uparrow$	HLA class II histocompatibility antigen, DR beta 4 chain	
TRINITY_DN154228_c0_g4	1.75	5.00	5.68E-07	0.002	$\uparrow$		
TRINITY_DN137109_c0_g3	1.38	4.97	6.70E-07	0.002	$\uparrow$		
TRINITY_DN142817_c1_g1	1.98	4.95	7.58E-07	0.002	$\uparrow$		
TRINITY_DN146802_c1_g6	2.05	4.82	1.44E-06	0.004	$\uparrow$		
TRINITY_DN168611_c5_g1	1.94	4.80	1.58E-06	0.004	$\uparrow$		
TRINITY_DN157075_c4_g4	1.84	4.75	2.05E-06	0.005	$\uparrow$		
TRINITY_DN141999_c0_g4	1.95	4.73	2.25E-06	0.005	$\uparrow$		
TRINITY_DN85147_c0_g2	1.99	4.67	3.06E-06	0.006	$\uparrow$		PREDICTED: carboxypeptidase N subunit 2-like [Poecilia formosa]
TRINITY_DN165613_c2_g5	1.89	4.66	3.13E-06	0.006	$\uparrow$		
TRINITY_DN172142_c2_g9	1.62	4.68	2.84E-06	0.006	$\uparrow$	Scavenger receptor cysteine-rich type 1 protein M130	
TRINITY_DN149278_c5_g1	1.51	4.65	3.37E-06	0.006	$\uparrow$		
TRINITY_DN148022_c3_g3	1.66	4.64	3.52E-06	0.006	$\uparrow$		
TRINITY_DN168513_c3_g1	1.85	4.50	6.84E-06	0.010	$\uparrow$		
TRINITY_DN153169_c0_g8	1.86	4.36	1.32E-05	0.017	$\uparrow$		
TRINITY_DN168556_c1_g4	1.85	4.37	1.24E-05	0.017	$\uparrow$	Endogenous retrovirus group 3 member 1 Env polyprotein	PREDICTED: uncharacterized protein LOC103181157 [Callorhinchus milii]
TRINITY_DN147941_c7_g2	1.58	4.36	1.27E-05	0.017	$\uparrow$		
TRINITY_DN172142_c2_g4	1.53	4.36	1.31E-05	0.017	$\uparrow$	Scavenger receptor cysteine-rich type 1 protein M130	
TRINITY_DN152661_c0_g10	1.74	4.31	1.66E-05	0.021	$\uparrow$		
TRINITY_DN160149_c1_g6	1.66	4.27	1.94E-05	0.023	$\uparrow$	Ig heavy chain V-I region HG3	PREDICTED: uncharacterized protein LOC101126730 [Gorilla gorilla gorilla]
TRINITY_DN115821_c0_g1	1.24	4.23	2.38E-05	0.026	$\uparrow$		
TRINITY_DN151416_c3_g2	1.51	4.20	2.63E-05	0.028	$\uparrow$		
TRINITY_DN158523_c0_g2	1.09	4.18	2.86E-05	0.030	$\uparrow$	Thymocyte nuclear protein 1	thymocyte nuclear protein 1 [Salmo salar]
TRINITY_DN150174_c4_g2	1.78	4.17	3.01E-05	0.031	$\uparrow$	Neprilysin	PREDICTED: neprilysin [Taeniopygia guttata]
TRINITY_DN163249_c0_g5	1.70	4.18	2.97E-05	0.031	$\uparrow$		
TRINITY_DN162522_c0_g3	1.06	4.13	3.60E-05	0.035	$\uparrow$		
TRINITY_DN131224_c0_g2	1.33	4.05	5.06E-05	0.044	$\uparrow$	Cathelicidin-5	cathelicidin 1 precursor [Oncorhynchus mykiss]
TRINITY_DN163556_c1_g10	1.69	4.03	5.69E-05	0.048	$\uparrow$		
TRINITY_DN164562_c0_g2	1.36	4.02	5.92E-05	0.050	$\uparrow$	Ferrochelatase, mitochondrial	PREDICTED: ferrochelatase, mitochondrial, partial [Poecilia formosa]
TRINITY_DN137702_c1_g8	-4.88	-12.80	1.56E-37	9.00E-33	$\downarrow$	Major histocompatibility complex class I-related gene protein	
TRINITY_DN162543_c1_g4	-3.11	-7.48	7.62E-14	2.20E-09	$\downarrow$	CD48 antigen	
TRINITY_DN158856_c0_g1	-2.76	-6.59	4.49E-11	5.64E-07	$\downarrow$		PREDICTED: stress response protein nst1-like [Poecilia formosa]

Gene ID						Protein name	Description
TRINITY_DN134666_c1_g1	-2.21	-5.64	1.72E-08	1.42E-04	↓	Uncharacterized protein KIAA0408	PREDICTED: uncharacterized protein KIAA0408 isoform X2 [Astyanax mexicanus]
TRINITY_DN172465_c6_g4	-2.14	-5.48	4.33E-08	2.78E-04	↓	Retrovirus-related Pol polyprotein from type-1 retrotransposable element R1	
TRINITY_DN172627_c9_g2	-1.89	-5.38	7.44E-08	4.31E-04	↓	Interferon-induced very large GTPase 1	PREDICTED: interferon-induced very large GTPase 1-like [Monodelphis domestica]
TRINITY_DN149496_c6_g5	-2.21	-5.27	1.36E-07	7.13E-04	↓		
TRINITY_DN164470_c0_g3	-1.69	-5.25	1.52E-07	7.34E-04	↓		
TRINITY_DN158064_c3_g2	-1.70	-5.23	1.72E-07	7.68E-04	↓	Early growth response protein 1	early growth response protein 1 [Danio rerio]
TRINITY_DN168080_c7_g1	-2.22	-5.19	2.08E-07	8.60E-04	↓		
TRINITY_DN155160_c1_g2	-2.15	-5.09	3.62E-07	0.001	↓		
TRINITY_DN170110_c2_g11	-2.10	-4.93	8.02E-07	0.002	↓	Periostin	
TRINITY_DN170542_c3_g2	-2.00	-4.78	1.72E-06	0.005	↓		
TRINITY_DN171747_c9_g6	-1.92	-4.68	2.92E-06	0.006	↓	Beta-galactoside alpha-2,6-sialyltransferase 2	
TRINITY_DN153446_c5_g2	-1.84	-4.67	3.01E-06	0.006	↓		
TRINITY_DN172635_c4_g1	-1.58	-4.65	3.26E-06	0.006	↓	Zinc finger BED domain-containing protein 1	PREDICTED: zinc finger BED domain-containing protein 1-like isoform X1 [Alligator sinensis]
TRINITY_DN142534_c2_g2	-1.90	-4.57	4.90E-06	0.009	↓		
TRINITY_DN171177_c5_g4	-1.91	-4.55	5.49E-06	0.009	↓	Cadherin EGF LAG seven-pass G-type receptor 2	PREDICTED: protocadherin Fat 4-like isoform X1 [Oryzias latipes]
TRINITY_DN164280_c3_g1	-1.87	-4.53	5.97E-06	0.009	↓	Beta-galactoside alpha-2,6-sialyltransferase 2	
TRINITY_DN164280_c3_g9	-1.65	-4.54	5.66E-06	0.009	↓	Beta-galactoside alpha-2,6-sialyltransferase 2	PREDICTED: beta-galactoside alpha-2,6-sialyltransferase 2-like [Astyanax mexicanus]
TRINITY_DN172484_c9_g1	-1.48	-4.53	5.91E-06	0.009	↓	Transposon Tf2-6 polyprotein	hypothetical protein CNBF1660 [Cryptococcus neoformans var. neoformans B-3501A]
TRINITY_DN165611_c2_g3	-1.86	-4.42	9.75E-06	0.014	↓	Major histocompatibility complex class I-related gene protein	
TRINITY_DN142915_c1_g1	-1.83	-4.40	1.08E-05	0.016	↓		PREDICTED: uncharacterized protein LOC101155681 [Oryzias latipes]
TRINITY_DN172071_c0_g11	-1.76	-4.34	1.45E-05	0.019	↓		
TRINITY_DN168913_c3_g2	-1.70	-4.27	1.92E-05	0.023	↓		
TRINITY_DN139249_c2_g10	-1.82	-4.26	2.01E-05	0.024	↓	GTPase IMAP family member 4	PREDICTED: GTPase IMAP family member 8-like, partial [Lepisosteus oculatus]
TRINITY_DN145333_c0_g3	-1.77	-4.26	2.04E-05	0.024	↓		
TRINITY_DN152192_c1_g1	-1.77	-4.25	2.17E-05	0.025	↓		
TRINITY_DN155888_c3_g2	-1.80	-4.23	2.39E-05	0.026	↓		
TRINITY_DN150042_c3_g3	-1.39	-4.14	3.48E-05	0.035	↓		PREDICTED: uncharacterized protein LOC100536567 [Danio rerio]
TRINITY_DN162962_c0_g2	-1.69	-4.13	3.61E-05	0.035	↓	Actin-related protein 2/3 complex subunit 1A	
TRINITY_DN146166_c1_g2	-1.72	-4.12	3.78E-05	0.036	↓		
TRINITY_DN157473_c0_g4	-1.69	-4.10	4.18E-05	0.038	↓		
TRINITY_DN154502_c3_g12	-1.74	-4.09	4.30E-05	0.038	↓		
TRINITY_DN169065_c0_g2	-1.46	-4.09	4.32E-05	0.038	↓	Early growth response protein 1	
Head kidney							
TRINITY_DN164373_c3_g4	4.46	7.17	7.66E-13	1.49E-08	↑		
TRINITY_DN160589_c0_g1	4.75	6.31	2.81E-10	2.34E-06	↑		PREDICTED: uncharacterized protein LOC101882462 [Danio rerio]
TRINITY_DN163744_c5_g3	3.96	6.09	1.11E-09	7.21E-06	↑	Retrovirus-related Pol polyprotein from type-1 retrotransposable element R2	PREDICTED: uncharacterized protein LOC580961 [Strongylocentrotus purpuratus]
TRINITY_DN142885_c0_g2	4.38	6.03	1.69E-09	9.85E-06	↑		
TRINITY_DN172445_c4_g1	4.04	5.61	2.02E-08	7.81E-05	↑		
TRINITY_DN146293_c0_g1	3.05	5.32	1.03E-07	3.34E-04	↑		
TRINITY_DN167068_c0_g1	3.26	5.27	1.33E-07	4.08E-04	↑	GRIN2-like protein	PREDICTED: GRIN2-like protein-like [Lepisosteus oculatus]
TRINITY_DN129930_c1_g1	2.55	5.23	1.65E-07	4.58E-04	↑	Protein CASC1	PREDICTED: protein CASC1 isoform X2 [Maylandia zebra]
TRINITY_DN142728_c3_g4	4.02	5.17	2.31E-07	5.85E-04	↑		
TRINITY_DN145690_c5_g7	3.87	5.14	2.71E-07	6.25E-04	↑		
TRINITY_DN155122_c3_g8	3.77	5.15	2.63E-07	6.25E-04	↑	Rho GDP-dissociation inhibitor 1	
TRINITY_DN142885_c0_g7	1.33	5.14	2.79E-07	6.25E-04	↑		
TRINITY_DN163556_c1_g10	4.00	5.11	3.26E-07	7.03E-04	↑		
TRINITY_DN146291_c0_g1	3.14	5.10	3.41E-07	7.09E-04	↑	Estrogen receptor beta	
TRINITY_DN164847_c1_g5	3.83	5.08	3.70E-07	7.43E-04	↑	HLA class II histocompatibility antigen, DR beta 4 chain	
TRINITY_DN135521_c0_g1	1.86	5.02	5.24E-07	9.55E-04	↑	Cytidine deaminase	Cytidine deaminase [Salmo salar]
TRINITY_DN172445_c4_g4	3.08	4.98	6.24E-07	0.001	↑		
TRINITY_DN163698_c2_g3	2.11	4.97	6.60E-07	0.001	↑	Collagen alpha-1(XXIV) chain	

TRINITY_DN125714_c0_g2	2.71	4.97	6.82E-07	0.001	↑	NEDD4-binding protein 3 homolog	PREDICTED: NEDD4-binding protein 3 homolog [Maylandia zebra]
TRINITY_DN137324_c2_g7	3.80	4.88	1.06E-06	0.002	↑		
TRINITY_DN168593_c4_g3	3.01	4.83	1.39E-06	0.002	↑	Estrogen receptor beta	
TRINITY_DN156999_c0_g1	3.13	4.77	1.82E-06	0.003	↑		PREDICTED: formin-like protein 5-like [Astyanax mexicanus]
TRINITY_DN129768_c0_g2	1.84	4.75	2.08E-06	0.003	↑	UBX domain-containing protein 11	PREDICTED: UBX domain-containing protein 11 isoform X1 [Danio rerio]
TRINITY_DN132913_c0_g2	3.46	4.68	2.93E-06	0.004	↑	FH2 domain-containing protein 1	PREDICTED: FH2 domain-containing protein 1 [Oreochromis niloticus]
TRINITY_DN170810_c0_g2	1.71	4.65	3.28E-06	0.004	↑	Multiple PDZ domain protein	
TRINITY_DN170462_c2_g14	1.69	4.66	3.24E-06	0.004	↑		
TRINITY_DN154760_c0_g1	3.46	4.64	3.43E-06	0.004	↑	Regulator of G-protein signaling 22	PREDICTED: regulator of G-protein signaling 22 isoform X2 [Xenopus (Silurana) tropicalis]
TRINITY_DN172445_c5_g1	2.83	4.64	3.43E-06	0.004	↑		
TRINITY_DN143340_c0_g1	3.25	4.63	3.61E-06	0.004	↑	Dynein heavy chain 7, axonemal	dynein heavy chain 7, axonemal [Bos taurus]
TRINITY_DN153332_c2_g1	3.07	4.57	4.84E-06	0.006	↑		PREDICTED: MANSC domain-containing protein 1 [Poecilia formosa]
TRINITY_DN136158_c0_g1	1.77	4.55	5.25E-06	0.006	↑	Receptor tyrosine-protein kinase erbB-3	PREDICTED: receptor tyrosine-protein kinase erbB-3-like isoform X1 [Maylandia zebra]
TRINITY_DN158138_c1_g1	2.05	4.55	5.28E-06	0.006	↑	Receptor-type tyrosine-protein phosphatase delta	PREDICTED: receptor-type tyrosine-protein phosphatase delta isoform X11 [Oreochromis niloticus]
TRINITY_DN159166_c1_g1	2.06	4.55	5.43E-06	0.006	↑	Nectin-4	PREDICTED: nectin-4 [Maylandia zebra]
TRINITY_DN139426_c1_g2	3.03	4.51	6.54E-06	0.007	↑	Biglycan	biglycan precursor [Danio rerio]
TRINITY_DN161979_c2_g3	3.30	4.50	6.74E-06	0.007	↑		
TRINITY_DN164354_c1_g3	3.47	4.50	6.79E-06	0.007	↑		
TRINITY_DN130640_c0_g2	2.70	4.46	8.11E-06	0.008	↑	EF-hand calcium-binding domain-containing protein 2	PREDICTED: EF-hand calcium-binding domain-containing protein 2-like [Lepisosteus oculatus]
TRINITY_DN129391_c1_g1	3.37	4.43	9.42E-06	0.008	↑		
TRINITY_DN151863_c1_g1	2.34	4.43	9.29E-06	0.008	↑	Collagen alpha-5(IV) chain	PREDICTED: collagen alpha-5(IV) chain [Astyanax mexicanus]
TRINITY_DN82942_c0_g1	3.23	4.43	9.42E-06	0.008	↑		
TRINITY_DN149909_c1_g3	1.88	4.45	8.75E-06	0.008	↑	Receptor-type tyrosine-protein phosphatase delta	PREDICTED: receptor-type tyrosine-protein phosphatase delta isoform 12 [Macaca mulatta]
TRINITY_DN157075_c4_g4	2.29	4.45	8.48E-06	0.008	↑		
TRINITY_DN161741_c3_g7	1.13	4.43	9.29E-06	0.008	↑		
TRINITY_DN122571_c0_g1	3.02	4.41	1.05E-05	0.009	↑	Dynein heavy chain 7, axonemal	PREDICTED: dynein heavy chain 7, axonemal [Macaca mulatta]
TRINITY_DN167020_c0_g5	2.09	4.41	1.05E-05	0.009	↑		
TRINITY_DN163104_c3_g2	2.43	4.37	1.23E-05	0.010	↑	Atrial natriuretic peptide receptor 1	
TRINITY_DN157378_c3_g1	2.13	4.38	1.21E-05	0.010	↑	Receptor-type tyrosine-protein phosphatase delta	PREDICTED: receptor-type tyrosine-protein phosphatase delta isoform 12 [Macaca mulatta]
TRINITY_DN161961_c4_g2	2.87	4.37	1.25E-05	0.010	↑	Carnitine O-palmitoyltransferase 1, muscle isoform	
TRINITY_DN139550_c0_g1	3.10	4.35	1.37E-05	0.011	↑	Uncharacterized protein C3orf67 homolog	PREDICTED: uncharacterized protein C3orf67 homolog isoform X1 [Macaca mulatta]
TRINITY_DN144825_c0_g1	1.63	4.35	1.37E-05	0.011	↑	Breast cancer anti-estrogen resistance protein 3	PREDICTED: SH2 domain-containing protein 3A isoform X1 [Oryzias latipes]
TRINITY_DN141576_c1_g3	1.65	4.33	1.46E-05	0.011	↑		
TRINITY_DN165538_c1_g1	3.15	4.32	1.53E-05	0.011	↑		
TRINITY_DN159222_c0_g1	1.76	4.32	1.57E-05	0.011	↑	Platelet-derived growth factor C	PREDICTED: platelet-derived growth factor C-like [Lepisosteus oculatus]
TRINITY_DN156373_c0_g1	2.66	4.31	1.64E-05	0.012	↑	PERQ amino acid-rich with GYF domain-containing protein 2	PREDICTED: PERQ amino acid-rich with GYF domain-containing protein 2-like [Lepisosteus oculatus]
TRINITY_DN151935_c1_g1	3.31	4.29	1.75E-05	0.012	↑		
TRINITY_DN160572_c1_g1	3.19	4.26	2.01E-05	0.013	↑		
TRINITY_DN151131_c2_g1	2.54	4.27	1.99E-05	0.013	↑		
TRINITY_DN149504_c0_g2	2.60	4.27	1.99E-05	0.013	↑	Heat-stable enterotoxin receptor	PREDICTED: heat-stable enterotoxin receptor-like [Lepisosteus oculatus]
TRINITY_DN157614_c0_g4	2.81	4.27	1.92E-05	0.013	↑		
TRINITY_DN153886_c5_g2	2.33	4.27	1.99E-05	0.013	↑	Epidermal growth factor receptor kinase substrate 8-like protein 1	PREDICTED: epidermal growth factor receptor kinase substrate 8-like protein 1 [Maylandia zebra]
TRINITY_DN171723_c2_g4	1.97	4.24	2.25E-05	0.015	↑	Receptor-type tyrosine-protein phosphatase delta	
TRINITY_DN171690_c2_g7	2.17	4.21	2.51E-05	0.016	↑	Podocalyxin	
TRINITY_DN162605_c1_g2	1.68	4.21	2.55E-05	0.016	↑	Wilms tumor protein	Wilms' tumor suppressor 1a [Oncorhynchus mykiss]

TRINITY_DN159754_c0_g1	2.75	4.21	2.60E-05	0.016	↑	Transient receptor potential cation channel subfamily M member 5	transient receptor potential cation channel subfamily M member 5 [Danio rerio]
TRINITY_DN148691_c0_g1	2.52	4.20	2.68E-05	0.016	↑	Receptor-interacting serine/threonine-protein kinase 4	PREDICTED: receptor-interacting serine/threonine-protein kinase 4 [Oreochromis niloticus]
TRINITY_DN148835_c0_g1	1.87	4.18	2.87E-05	0.017	↑		
TRINITY_DN169331_c3_g2	1.40	4.17	3.00E-05	0.018	↑	1-phosphatidylinositol 4,5-bisphosphate phosphodiesterase delta	PREDICTED: 1-phosphatidylinositol 4,5-bisphosphate phosphodiesterase delta-1 isoform X1 [Taeniopygia guttata]
TRINITY_DN163384_c3_g2	2.75	4.16	3.14E-05	0.018	↑		
TRINITY_DN137274_c3_g4	2.42	4.16	3.19E-05	0.018	↑		
TRINITY_DN164726_c4_g1	1.50	4.15	3.26E-05	0.018	↑	Secretory phospholipase A2 receptor	PREDICTED: secretory phospholipase A2 receptor-like [Lepisosteus oculatus]
TRINITY_DN159521_c2_g1	2.23	4.15	3.38E-05	0.019	↑		
TRINITY_DN146016_c7_g1	2.77	4.14	3.44E-05	0.019	↑		
TRINITY_DN159940_c0_g1	2.02	4.14	3.47E-05	0.019	↑		PREDICTED: multiple PDZ domain protein-like [Lepisosteus oculatus]
TRINITY_DN172377_c0_g1	1.75	4.13	3.63E-05	0.019	↑	Extracellular matrix protein FRAS1	extracellular matrix protein FRAS1 [Danio rerio]
TRINITY_DN147651_c0_g2	3.19	4.12	3.75E-05	0.020	↑		
TRINITY_DN89013_c0_g1	2.96	4.12	3.83E-05	0.020	↑	Electrogenic sodium bicarbonate cotransporter 1	PREDICTED: electrogenic sodium bicarbonate cotransporter 1-like [Lepisosteus oculatus]
TRINITY_DN164224_c0_g1	1.47	4.12	3.84E-05	0.020	↑	Sortilin-related receptor	PREDICTED: sortilin-related receptor [Monodelphis domestica]
TRINITY_DN170291_c0_g1	2.75	4.11	3.96E-05	0.020	↑	Protein prune homolog 2	PREDICTED: protein prune homolog 2-like [Lepisosteus oculatus]
TRINITY_DN163108_c0_g1	2.04	4.10	4.21E-05	0.021	↑		
TRINITY_DN159545_c1_g1	1.52	4.09	4.27E-05	0.021	↑		
TRINITY_DN157378_c3_g2	2.04	4.09	4.35E-05	0.021	↑	Receptor-type tyrosine-protein phosphatase delta	
TRINITY_DN162200_c1_g1	2.02	4.07	4.79E-05	0.023	↑		
TRINITY_DN160149_c1_g6	1.65	4.05	5.02E-05	0.023	↑	Ig heavy chain V-I region HG3	PREDICTED: uncharacterized protein LOC101126730 [Gorilla gorilla gorilla]
TRINITY_DN157434_c0_g2	2.10	4.06	4.95E-05	0.023	↑	Centrosomal protein cep57l1	
TRINITY_DN128669_c0_g2	1.90	4.05	5.19E-05	0.023	↑	Serine/threonine-protein kinase pim-3	PREDICTED: serine/threonine-protein kinase pim-3-like [Astyanax mexicanus]
TRINITY_DN161441_c1_g1	2.43	4.05	5.17E-05	0.023	↑		
TRINITY_DN156836_c3_g2	1.42	4.05	5.15E-05	0.023	↑	Receptor-type tyrosine-protein phosphatase delta	
TRINITY_DN146877_c0_g1	1.97	4.05	5.10E-05	0.023	↑	Microtubule-associated tumor suppressor candidate 2	PREDICTED: microtubule-associated tumor suppressor candidate 2-like [Xiphophorus maculatus]
TRINITY_DN138538_c0_g1	2.81	4.04	5.28E-05	0.023	↑	Disabled homolog 1	PREDICTED: disabled homolog 1-like [Lepisosteus oculatus]
TRINITY_DN131031_c0_g1	2.33	4.04	5.27E-05	0.023	↑		PREDICTED: uncharacterized protein LOC100893945 [Strongylocentrotus purpuratus]
TRINITY_DN114237_c0_g2	3.11	4.04	5.41E-05	0.024	↑		
TRINITY_DN171512_c0_g1	2.44	4.04	5.45E-05	0.024	↑	Desmoplakin	PREDICTED: desmoplakin isoform X1 [Danio rerio]
TRINITY_DN168842_c1_g8	1.57	4.02	5.82E-05	0.025	↑		
TRINITY_DN152103_c1_g1	1.65	4.01	6.05E-05	0.026	↑	Tripartite motif-containing protein 16	finTRIM family, member 82 [Danio rerio]
TRINITY_DN138430_c0_g2	2.56	4.00	6.41E-05	0.026	↑		
TRINITY_DN157475_c0_g2	3.03	4.00	6.45E-05	0.026	↑		hypothetical protein AOL_s00080g372 [Arthrobotrys oligospora ATCC 24927]
TRINITY_DN147571_c3_g13	3.18	3.98	6.90E-05	0.027	↑		
TRINITY_DN164491_c3_g1	2.17	3.98	7.00E-05	0.027	↑		
TRINITY_DN153657_c2_g1	2.15	3.97	7.13E-05	0.028	↑	Collagen alpha-1(IV) chain	PREDICTED: collagen alpha-5(IV) chain-like isoform X1 [Python bivittatus]
TRINITY_DN167740_c2_g1	2.41	3.96	7.36E-05	0.028	↑	Synaptotagmin-like protein 2	PREDICTED: synaptotagmin-like protein 2-like [Lepisosteus oculatus]
TRINITY_DN163489_c1_g1	2.48	3.96	7.39E-05	0.028	↑		
TRINITY_DN159004_c2_g1	2.80	3.96	7.57E-05	0.029	↑	LINE-1 retrotransposable element ORF2 protein	PREDICTED: uncharacterized protein LOC100333395 [Danio rerio]
TRINITY_DN144690_c2_g1	2.61	3.96	7.65E-05	0.029	↑		
TRINITY_DN171079_c3_g4	2.43	3.95	7.80E-05	0.029	↑	Rho guanine nucleotide exchange factor 28	PREDICTED: rho guanine nucleotide exchange factor 28-like [Lepisosteus oculatus]
TRINITY_DN161130_c4_g1	2.54	3.95	7.77E-05	0.029	↑	Estrogen receptor beta	

TRINITY_DN133663_c0_g1	2.67	3.94	7.99E-05	0.029	↑	Microtubule-actin cross-linking factor 1, isoforms 1/2/3/5	PREDICTED: microtubule-actin cross-linking factor 1, isoform 4-like [Macaca mulatta]
TRINITY_DN127233_c1_g1	2.91	3.94	7.98E-05	0.029	↑	Sex hormone-binding globulin	sex hormone-binding globulin precursor [Danio rerio]
TRINITY_DN145305_c0_g1	2.94	3.94	8.06E-05	0.029	↑		
TRINITY_DN124704_c0_g1	2.64	3.94	8.25E-05	0.030	↑	Fermitin family homolog 1	PREDICTED: fermitin family homolog 1 [Equus caballus]
TRINITY_DN153536_c0_g2	3.01	3.94	8.28E-05	0.030	↑		
TRINITY_DN169894_c1_g1	2.28	3.92	8.77E-05	0.031	↑	Protein Shroom1	PREDICTED: protein Shroom1-like [Lepisosteus oculatus]
TRINITY_DN145654_c0_g1	2.48	3.92	8.70E-05	0.031	↑	Aquaporin-5	PREDICTED: lens fiber major intrinsic protein-like [Lepisosteus oculatus]
TRINITY_DN158659_c0_g1	2.55	3.92	8.79E-05	0.031	↑		PREDICTED: uncharacterized protein LOC102313737 [Haplochromis burtoni]
TRINITY_DN161783_c2_g6	2.84	3.92	8.98E-05	0.031	↑		
TRINITY_DN117671_c0_g1	3.01	3.91	9.06E-05	0.031	↑		
TRINITY_DN147394_c0_g1	1.62	3.91	9.23E-05	0.032	↑		
TRINITY_DN127456_c0_g1	2.65	3.90	9.46E-05	0.032	↑	Microtubule-actin cross-linking factor 1	PREDICTED: microtubule-actin cross-linking factor 1, isoforms 1/2/3/5 isoform X19 [Astyanax mexicanus]
TRINITY_DN126333_c0_g3	1.74	3.90	9.46E-05	0.032	↑		
TRINITY_DN168556_c1_g4	2.37	3.89	9.88E-05	0.033	↑	Endogenous retrovirus group 3 member 1 Env polyprotein	PREDICTED: uncharacterized protein LOC103181157 [Callorhinchus milii]
TRINITY_DN143658_c5_g1	1.18	3.88	1.04E-04	0.034	↑	Arf-GAP with coiled-coil, ANK repeat and PH domain-containing protein 3	arf-GAP with coiled-coil, ANK repeat and PH domain-containing protein 3 [Danio rerio]
TRINITY_DN142728_c3_g3	3.01	3.88	1.04E-04	0.034	↑		
TRINITY_DN166483_c1_g3	1.90	3.88	1.04E-04	0.034	↑		
TRINITY_DN118131_c0_g1	2.67	3.88	1.06E-04	0.034	↑		
TRINITY_DN119036_c0_g1	2.83	3.87	1.08E-04	0.035	↑	Ras association domain-containing protein 5	PREDICTED: ras association domain-containing protein 3 isoform X1 [Danio rerio]
TRINITY_DN108108_c0_g1	2.63	3.87	1.09E-04	0.035	↑		
TRINITY_DN172092_c3_g5	2.14	3.86	1.12E-04	0.035	↑	Pleckstrin homology domain-containing family A member 6	
TRINITY_DN129545_c0_g1	2.24	3.86	1.14E-04	0.035	↑	Calcium uptake protein 3, mitochondrial	PREDICTED: calcium uptake protein 3, mitochondrial-like isoform X5 [Poecilia formosa]
TRINITY_DN141576_c2_g1	1.25	3.86	1.13E-04	0.035	↑		
TRINITY_DN156529_c4_g7	2.52	3.86	1.14E-04	0.035	↑	Histidine ammonia-lyase	
TRINITY_DN169063_c1_g1	1.49	3.85	1.16E-04	0.036	↑	Retrotransposon-derived protein PEG10	PREDICTED: retrotransposon-like protein 1-like [Lepisosteus oculatus]
TRINITY_DN142758_c0_g1	2.36	3.85	1.16E-04	0.036	↑		PREDICTED: basement membrane-specific heparan sulfate proteoglycan core protein-like, partial [Astyanax mexicanus]
TRINITY_DN166500_c2_g1	2.04	3.85	1.19E-04	0.036	↑		
TRINITY_DN170810_c0_g4	1.57	3.84	1.21E-04	0.036	↑	Multiple PDZ domain protein	multiple PDZ domain protein [Bos taurus]
TRINITY_DN150786_c3_g5	2.27	3.84	1.21E-04	0.036	↑		
TRINITY_DN168714_c1_g2	1.73	3.84	1.25E-04	0.037	↑	Feline leukemia virus subgroup C receptor-related protein 1	
TRINITY_DN156129_c0_g2	2.17	3.83	1.26E-04	0.037	↑	UDP-glucuronosyltransferase 1	UDP glucuronosyltransferase 1 family, polypeptide B7 precursor [Danio rerio]
TRINITY_DN167796_c0_g20	2.21	3.83	1.26E-04	0.037	↑		
TRINITY_DN153819_c1_g1	1.46	3.83	1.29E-04	0.037	↑		
TRINITY_DN8610_c0_g1	2.86	3.83	1.29E-04	0.037	↑		
TRINITY_DN156187_c0_g4	1.25	3.83	1.30E-04	0.037	↑		
TRINITY_DN167532_c0_g2	1.13	3.83	1.29E-04	0.037	↑		
TRINITY_DN158294_c1_g1	2.37	3.83	1.29E-04	0.037	↑		
TRINITY_DN143340_c1_g1	2.65	3.83	1.29E-04	0.037	↑	Dynein heavy chain 10, axonemal	
TRINITY_DN131751_c0_g1	2.66	3.82	1.33E-04	0.037	↑		
TRINITY_DN150997_c0_g2	2.63	3.82	1.33E-04	0.037	↑		
TRINITY_DN159523_c0_g1	1.80	3.82	1.33E-04	0.037	↑	Calmodulin-binding transcription activator 1	PREDICTED: calmodulin-binding transcription activator 1-like [Lepisosteus oculatus]
TRINITY_DN138685_c0_g2	2.47	3.82	1.35E-04	0.037	↑	Hepatocyte nuclear factor 3-beta	hepatocyte nuclear factor 3-gamma [Danio rerio]
TRINITY_DN154878_c1_g1	2.00	3.82	1.35E-04	0.037	↑	Transmembrane protein 82	PREDICTED: transmembrane protein 82-like isoform X1 [Astyanax mexicanus]
TRINITY_DN158955_c2_g7	2.57	3.81	1.37E-04	0.038	↑	Pleckstrin homology domain-containing family A member 6	PREDICTED: pleckstrin homology domain-containing family A member 6 isoform X8 [Astyanax mexicanus]
TRINITY_DN141962_c0_g2	2.03	3.81	1.40E-04	0.038	↑		PREDICTED: location of vulva defective 1-like isoform X1 [Latimeria chalumnae]

Transcript ID						Gene name	Description
TRINITY_DN155287_c2_g2	1.26	3.80	1.43E-04	0.039	↑		
TRINITY_DN151813_c0_g1	2.80	3.80	1.44E-04	0.039	↑	WD repeat-containing protein 38	PREDICTED: vegetative incompatibility protein HET-E-1-like [Astyanax mexicanus]
TRINITY_DN114883_c0_g1	2.50	3.80	1.46E-04	0.039	↑	Sperm-associated antigen 17	sperm-associated antigen 17 [Xenopus (Silurana) tropicalis]
TRINITY_DN157557_c2_g1	2.33	3.80	1.46E-04	0.039	↑	PR domain zinc finger protein 16	
TRINITY_DN161854_c1_g1	1.17	3.80	1.45E-04	0.039	↑	Filamin-A-interacting protein 1	PREDICTED: filamin-A-interacting protein 1 isoform X1 [Equus caballus]
TRINITY_DN153662_c1_g1	2.64	3.79	1.48E-04	0.039	↑	Claudin-8	PREDICTED: claudin-8-like [Astyanax mexicanus]
TRINITY_DN171566_c0_g1	1.24	3.79	1.49E-04	0.039	↑	RING finger protein 24	PREDICTED: RING finger protein 24 [Columba livia]
TRINITY_DN141782_c0_g1	1.88	3.80	1.47E-04	0.039	↑	Reversion-inducing cysteine-rich protein with Kazal motifs	PREDICTED: reversion-inducing cysteine-rich protein with Kazal motifs-like [Lepisosteus oculatus]
TRINITY_DN150997_c0_g3	2.65	3.79	1.51E-04	0.039	↑		
TRINITY_DN154771_c0_g4	2.66	3.79	1.51E-04	0.039	↑		
TRINITY_DN140976_c1_g1	1.42	3.78	1.55E-04	0.040	↑	Riboflavin transporter 2	riboflavin transporter 2 [Salmo salar]
TRINITY_DN168692_c0_g7	2.45	3.78	1.57E-04	0.040	↑		
TRINITY_DN156385_c1_g1	2.48	3.77	1.64E-04	0.041	↑		PREDICTED: probable GPI-anchored adhesin-like protein PGA55-like isoform X1 [Astyanax mexicanus]
TRINITY_DN149035_c4_g1	1.71	3.77	1.65E-04	0.041	↑	Beta-1,4 N-acetylgalactosaminyltransferase 2	PREDICTED: beta-1,4 N-acetylgalactosaminyltransferase 2-like [Lepisosteus oculatus]
TRINITY_DN166950_c1_g1	2.06	3.76	1.70E-04	0.042	↑	Collagen alpha-1(IV) chain	PREDICTED: collagen alpha-5(IV) chain isoform X1 [Danio rerio]
TRINITY_DN160244_c1_g1	1.16	3.75	1.74E-04	0.043	↑	Protocadherin-1	protocadherin-1 isoform 2 precursor [Homo sapiens]
TRINITY_DN118678_c1_g1	2.52	3.75	1.76E-04	0.043	↑		
TRINITY_DN140127_c3_g1	2.05	3.75	1.77E-04	0.043	↑		
TRINITY_DN142125_c0_g1	1.91	3.75	1.79E-04	0.043	↑	C-myc promoter-binding protein	C-myc promoter-binding protein [Danio rerio]
TRINITY_DN165320_c4_g1	2.36	3.74	1.81E-04	0.044	↑	Protein kinase C delta type	PREDICTED: protein kinase C delta type [Anolis carolinensis]
TRINITY_DN166483_c1_g4	1.05	3.73	1.90E-04	0.045	↑		
TRINITY_DN138257_c0_g1	1.98	3.73	1.90E-04	0.045	↑		
TRINITY_DN169152_c1_g2	1.29	3.73	1.90E-04	0.045	↑	Sortilin-related receptor	
TRINITY_DN134522_c1_g1	2.39	3.73	1.90E-04	0.045	↑	Biotinidase	biotinidase precursor [Danio rerio]
TRINITY_DN163384_c3_g11	2.61	3.73	1.94E-04	0.046	↑	MANSC domain-containing protein 1	
TRINITY_DN145200_c1_g2	2.71	3.73	1.95E-04	0.046	↑	Desmoglein-1	
TRINITY_DN172092_c3_g1	2.18	3.72	1.97E-04	0.046	↑		
TRINITY_DN151151_c1_g4	1.51	3.72	1.98E-04	0.046	↑		
TRINITY_DN160367_c1_g2	1.67	3.72	1.99E-04	0.046	↑	Multiple PDZ domain protein	PREDICTED: multiple PDZ domain protein isoform X6 [Oryctolagus cuniculus]
TRINITY_DN142233_c0_g1	1.58	3.72	2.01E-04	0.046	↑	Secretory phospholipase A2 receptor	PREDICTED: secretory phospholipase A2 receptor [Poecilia formosa]
TRINITY_DN161659_c3_g4	1.29	3.71	2.04E-04	0.046	↑		
TRINITY_DN169159_c1_g1	1.98	3.71	2.04E-04	0.046	↑		
TRINITY_DN146606_c2_g1	2.96	3.71	2.05E-04	0.046	↑		
TRINITY_DN150087_c0_g1	1.42	3.71	2.04E-04	0.046	↑		
TRINITY_DN170841_c1_g1	2.14	3.71	2.08E-04	0.047	↑		
TRINITY_DN163698_c2_g1	1.58	3.70	2.12E-04	0.047	↑	Collagen alpha-1(XXIV) chain	PREDICTED: collagen alpha-1(XXIV) chain-like [Lepisosteus oculatus]
TRINITY_DN73500_c0_g1	2.71	3.70	2.12E-04	0.047	↑		
TRINITY_DN170036_c1_g2	2.07	3.70	2.14E-04	0.047	↑		
TRINITY_DN156175_c0_g1	2.49	3.69	2.24E-04	0.048	↑		
TRINITY_DN167941_c0_g4	1.14	3.70	2.18E-04	0.048	↑		
TRINITY_DN166241_c2_g1	1.89	3.69	2.21E-04	0.048	↑	Homeodomain-interacting protein kinase 2	PREDICTED: homeodomain-interacting protein kinase 2 isoform X3 [Maylandia zebra]
TRINITY_DN169903_c2_g2	2.70	3.70	2.17E-04	0.048	↑		
TRINITY_DN158655_c3_g2	1.19	3.69	2.27E-04	0.048	↑		
TRINITY_DN155119_c1_g1	2.37	3.68	2.36E-04	0.049	↑	Glycerophosphodiester phosphodiesterase domain-containing protein 5	PREDICTED: glycerophosphoinositol inositolphosphodiesterase GDPD2 isoform X1 [Astyanax mexicanus]
TRINITY_DN172465_c4_g1	2.52	3.68	2.37E-04	0.049	↑		
TRINITY_DN159491_c0_g2	1.59	3.67	2.40E-04	0.050	↑	7-alpha-hydroxycholest-4-en-3-one 12-alpha-hydroxylase	cytochrome P450, family 8, subfamily B, polypeptide 2 [Danio rerio]
TRINITY_DN137702_c1_g10	-5.33	-7.64	2.14E-14	6.24E-10	↓	Major histocompatibility complex class I-related gene protein	
TRINITY_DN137702_c1_g8	-5.28	-7.68	1.63E-14	6.24E-10	↓	Major histocompatibility complex class I-related gene protein	
TRINITY_DN155160_c1_g2	-3.93	-6.90	5.05E-12	7.35E-08	↓		
TRINITY_DN141783_c3_g5	-4.84	-6.33	2.43E-10	2.34E-06	↓		
TRINITY_DN171219_c1_g6	-3.52	-6.32	2.62E-10	2.34E-06	↓		

Gene	Log2FC	Wald stat	p-value	adj p-value	Direction	Annotation	Annotation
TRINITY_DN172357_c4_g1	-4.48	-6.27	3.54E-10	2.57E-06	↓		
TRINITY_DN166778_c1_g2	-2.33	-6.00	1.98E-09	1.05E-05	↓		
TRINITY_DN155813_c1_g1	-2.58	-5.89	3.80E-09	1.84E-05	↓	BTB/POZ domain-containing protein 17	PREDICTED: BTB/POZ domain-containing protein 17-like [Latimeria chalumnae]
TRINITY_DN168702_c3_g7	-2.25	-5.85	5.05E-09	2.27E-05	↓		
TRINITY_DN118037_c0_g1	-4.47	-5.76	8.32E-09	3.46E-05	↓		
TRINITY_DN143384_c0_g2	-3.40	-5.60	2.15E-08	7.81E-05	↓		
TRINITY_DN170991_c4_g8	-4.22	-5.45	5.04E-08	1.73E-04	↓		
TRINITY_DN147195_c0_g2	-3.99	-5.26	1.47E-07	4.28E-04	↓		
TRINITY_DN158618_c1_g2	-1.02	-5.18	2.27E-07	5.85E-04	↓		membrane protein ORF71 [Anguillid herpesvirus 1]
TRINITY_DN164000_c0_g3	-2.21	-5.07	4.00E-07	7.74E-04	↓		
TRINITY_DN168857_c2_g5	-2.86	-4.91	9.28E-07	0.002	↓	GTPase IMAP family member 4	
TRINITY_DN162543_c1_g4	-3.61	-4.81	1.51E-06	0.002	↓	CD48 antigen	
TRINITY_DN172220_c8_g6	-1.39	-4.53	6.03E-06	0.007	↓		
TRINITY_DN169367_c0_g2	-1.01	-4.53	6.00E-06	0.007	↓		
TRINITY_DN160915_c1_g1	-2.90	-4.44	8.80E-06	0.008	↓		PREDICTED: uncharacterized protein LOC103032205 [Astyanax mexicanus]
TRINITY_DN141485_c1_g9	-2.74	-4.44	8.91E-06	0.008	↓		
TRINITY_DN167602_c0_g6	-1.72	-4.38	1.21E-05	0.010	↓		
TRINITY_DN152464_c2_g16	-3.15	-4.34	1.43E-05	0.011	↓	Granulins	
TRINITY_DN158192_c0_g1	-2.38	-4.34	1.43E-05	0.011	↓		PREDICTED: uncharacterized protein C2orf16-like [Poecilia formosa]
TRINITY_DN151272_c3_g7	-2.29	-4.21	2.52E-05	0.016	↓		
TRINITY_DN168771_c3_g12	-3.06	-4.18	2.92E-05	0.017	↓		
TRINITY_DN120453_c1_g2	-1.90	-4.16	3.14E-05	0.018	↓	Ig heavy chain V-III region VH26	PREDICTED: uncharacterized protein LOC101132759 [Gorilla gorilla gorilla]
TRINITY_DN167974_c0_g4	-1.47	-4.16	3.16E-05	0.018	↓		
TRINITY_DN170991_c4_g12	-3.20	-4.14	3.41E-05	0.019	↓		PREDICTED: uncharacterized protein LOC103172695 [Callorhinchus milii]
TRINITY_DN144819_c0_g4	-1.60	-4.13	3.64E-05	0.019	↓	P-selectin	
TRINITY_DN168913_c3_g2	-2.13	-4.08	4.42E-05	0.021	↓		
TRINITY_DN144089_c4_g6	-1.53	-4.09	4.39E-05	0.021	↓	Ig lambda chain V-V region DEL	
TRINITY_DN161509_c2_g4	-3.12	-4.06	4.95E-05	0.023	↓	Steroid 17-alpha-hydroxylase/17,20 lyase	
TRINITY_DN169753_c9_g14	-2.92	-4.03	5.60E-05	0.024	↓		
TRINITY_DN168610_c4_g1	-2.44	-4.01	6.15E-05	0.026	↓		
TRINITY_DN172635_c4_g1	-1.92	-4.01	6.15E-05	0.026	↓	Zinc finger BED domain-containing protein 1	PREDICTED: zinc finger BED domain-containing protein 1-like isoform X1 [Alligator sinensis]
TRINITY_DN167602_c0_g9	-1.23	-4.01	6.20E-05	0.026	↓		
TRINITY_DN128201_c0_g3	-3.05	-4.00	6.45E-05	0.026	↓	CMRF35-like molecule	
TRINITY_DN168074_c3_g4	-2.40	-4.00	6.47E-05	0.026	↓		
TRINITY_DN139287_c8_g14	-1.12	-3.99	6.52E-05	0.026	↓		
TRINITY_DN166578_c1_g6	-2.72	-3.98	6.86E-05	0.027	↓		
TRINITY_DN151727_c0_g1	-1.34	-3.98	6.96E-05	0.027	↓		
TRINITY_DN164265_c0_g2	-1.05	-3.95	7.87E-05	0.029	↓		
TRINITY_DN140173_c6_g1	-1.78	-3.93	8.34E-05	0.030	↓		
TRINITY_DN165801_c3_g2	-1.48	-3.89	9.86E-05	0.033	↓		
TRINITY_DN152706_c1_g13	-2.58	-3.89	1.02E-04	0.034	↓	GTPase IMAP family member 4	
TRINITY_DN137919_c0_g1	-2.48	-3.87	1.09E-04	0.035	↓		
TRINITY_DN156202_c2_g3	-2.91	-3.85	1.18E-04	0.036	↓	IgGFc-binding protein	PREDICTED: IgGFc-binding protein-like [Lepisosteus oculatus]
TRINITY_DN154217_c0_g11	-1.59	-3.83	1.26E-04	0.037	↓		
TRINITY_DN142651_c0_g2	-1.89	-3.82	1.33E-04	0.037	↓		
TRINITY_DN147923_c2_g2	-2.66	-3.81	1.38E-04	0.038	↓	Myosin-11	
TRINITY_DN165611_c2_g3	-2.35	-3.80	1.42E-04	0.038	↓	Major histocompatibility complex class I-related gene protein	
TRINITY_DN168771_c3_g3	-2.90	-3.78	1.57E-04	0.040	↓	Ig kappa chain V-V region HP 91A3	
TRINITY_DN150396_c1_g9	-3.01	-3.78	1.58E-04	0.040	↓		
TRINITY_DN164893_c1_g8	-2.17	-3.78	1.60E-04	0.040	↓	Myelin-oligodendrocyte glycoprotein	PREDICTED: HERV-H LTR-associating protein 2-like [Lepisosteus oculatus]
TRINITY_DN151002_c1_g1	-1.55	-3.74	1.85E-04	0.044	↓	Membrane-spanning 4-domains subfamily A member 12	PREDICTED: high affinity immunoglobulin epsilon receptor subunit beta-like isoform X1 [Panthera tigris altaica]
TRINITY_DN142646_c0_g2	-2.85	-3.69	2.20E-04	0.048	↓		
TRINITY_DN159589_c1_g3	-2.83	-3.69	2.20E-04	0.048	↓		
TRINITY_DN159103_c2_g10	-2.63	-3.69	2.23E-04	0.048	↓		
TRINITY_DN168857_c2_g10	-2.68	-3.69	2.27E-04	0.048	↓		
TRINITY_DN157162_c0_g5	-1.82	-3.68	2.30E-04	0.049	↓		

Annotations, where available, were obtained by blastx searches against UniProt and RefSeq databases. Genes that were differentially expressed in both tissues were shaded in gray. Log2FC is the log2 fold change between expression in infected and control individuals, Wald stat is the Wald statistic calculated by DESeq2, and Direction indicates up-regulation (↑) or down-regulation (↓) in infected individuals.

Table SI.2 List of all enriched GO terms in the spleen. Terms that were also enriched in the head kidney are shaded in gray.

Category	GO term (biological process)	Expected	Count	Size	Direction	p-value
GO:0006953	acute-phase response	0	2	119	↑	<0.001
GO:0001788	antibody-dependent cellular cytotoxicity	0	1	3	↑	<0.001
GO:0046449	creatinine metabolic process	0	1	4	↑	0.001
GO:0001805	positive regulation of type III hypersensitivity	0	1	6	↑	0.001
GO:0019884	antigen processing and presentation of exogenous antigen	0	2	360	↑	0.001
GO:0006954	inflammatory response	0	3	1754	↑	0.001
GO:0032672	regulation of interleukin-3 production	0	1	11	↑	0.001
GO:0002003	angiotensin maturation	0	1	11	↑	0.001
GO:0045425	positive regulation of granulocyte macrophage colony-stimulating factor biosynthetic process	0	1	11	↑	0.001
GO:0045401	positive regulation of interleukin-3 biosynthetic process	0	1	11	↑	0.001
GO:0071493	cellular response to UV-B	0	1	13	↑	0.002
GO:0071492	cellular response to UV-A	0	1	14	↑	0.002
GO:0001798	positive regulation of type IIa hypersensitivity	0	1	14	↑	0.002
GO:0002892	regulation of type II hypersensitivity	0	1	14	↑	0.002
GO:0048002	antigen processing and presentation of peptide antigen	0	2	543	↑	0.002
GO:0042742	defense response to bacterium	0	2	550	↑	0.002
GO:0001812	positive regulation of type I hypersensitivity	0	1	21	↑	0.003
GO:0090399	replicative senescence	0	1	21	↑	0.003
GO:0039657	suppression by virus of host gene expression	0	1	22	↑	0.003
GO:0060177	regulation of angiotensin metabolic process	0	1	23	↑	0.003
GO:0001820	serotonin secretion	0	1	24	↑	0.003
GO:0032645	regulation of granulocyte macrophage colony-stimulating factor production	0	1	25	↑	0.003
GO:0002866	positive regulation of acute inflammatory response to antigenic stimulus	0	1	33	↑	0.004
GO:0002524	hypersensitivity	0	1	39	↑	0.005
GO:0043306	positive regulation of mast cell degranulation	0	1	44	↑	0.006
GO:0002696	positive regulation of leukocyte activation	0	2	962	↑	0.006
GO:0002861	regulation of inflammatory response to antigenic stimulus	0	1	49	↑	0.006
GO:0001991	regulation of systemic arterial blood pressure by circulatory renin-angiotensin	0	1	53	↑	0.007
GO:0075732	viral penetration into host nucleus	0	1	55	↑	0.007
GO:0006898	receptor-mediated endocytosis	0	2	1076	↑	0.008
GO:0019370	leukotriene biosynthetic process	0	1	62	↑	0.008
GO:0042590	antigen processing and presentation of exogenous peptide antigen via MHC class I	0	1	76	↑	0.010
GO:0097503	sialylation	0	3	86	↓	<0.001
GO:0009311	oligosaccharide metabolic process	0	3	216	↓	<0.001
GO:0018279	protein N-linked glycosylation via asparagine	0	3	357	↓	<0.001
GO:0048703	embryonic viscerocranium morphogenesis	0	2	85	↓	<0.001
GO:0060059	embryonic retina morphogenesis in camera-type eye	0	2	99	↓	<0.001
GO:0006486	protein glycosylation	0	3	1009	↓	0.001
GO:0070085	glycosylation	0	3	1025	↓	0.001
GO:0002474	antigen processing and presentation of peptide antigen via MHC class I	0	2	342	↓	0.002
GO:0009100	glycoprotein metabolic process	0	3	1463	↓	0.002
GO:0048704	embryonic skeletal system morphogenesis	0	2	397	↓	0.003
GO:0021999	neural plate anterior/posterior regionalization	0	1	20	↓	0.004
GO:1904209	positive regulation of chemokine (C-C motif) ligand 2 secretion	0	1	22	↓	0.004
GO:1990523	bone regeneration	0	1	22	↓	0.004
GO:0033326	cerebrospinal fluid secretion	0	1	23	↓	0.004
GO:0048593	camera-type eye morphogenesis	0	2	612	↓	0.006
GO:0019882	antigen processing and presentation	0	2	651	↓	0.007
GO:2000341	regulation of chemokine (C-X-C motif) ligand 2 production	0	1	37	↓	0.007
GO:0060041	retina development in camera-type eye	0	2	716	↓	0.008
GO:1901137	carbohydrate derivative biosynthetic process	0	3	2385	↓	0.009
GO:0090196	regulation of chemokine secretion	0	1	52	↓	0.01

Terms that were also enriched in the head kidney are shaded in gray.

Table SI.3 List of all enriched GO terms in the head kidney. Terms that were also enriched in the spleen are shaded in gray.

Category	GO term (biological process)	Expected	Count	Size	Direction	p-value
GO:0097105	presynaptic membrane assembly	0	6	89	↑	<0.001
GO:0007157	heterophilic cell-cell adhesion via plasma membrane cell adhesion molecules	0	7	238	↑	<0.001
GO:0050775	positive regulation of dendrite morphogenesis	0	6	162	↑	<0.001
GO:0044091	membrane biogenesis	0	6	298	↑	<0.001
GO:0022610	biological adhesion	9	23	6659	↑	<0.001
GO:1900139	negative regulation of arachidonic acid secretion	0	2	4	↑	<0.001
GO:1900138	negative regulation of phospholipase A2 activity	0	2	6	↑	<0.001
GO:0007185	transmembrane receptor protein tyrosine phosphatase signaling pathway	0	3	46	↑	<0.001
GO:0046426	negative regulation of JAK-STAT cascade	0	4	163	↑	<0.001
GO:0098609	cell-cell adhesion	5	15	3724	↑	<0.001
GO:0010976	positive regulation of neuron projection development	1	8	1128	↑	<0.001
GO:0048856	anatomical structure development	27	43	20966	↑	<0.001
GO:0090238	positive regulation of arachidonic acid secretion	0	2	13	↑	<0.001
GO:0050808	synapse organization	2	8	1210	↑	<0.001
GO:0090403	oxidative stress-induced premature senescence	0	2	16	↑	<0.001
GO:0086073	bundle of His cell-Purkinje myocyte adhesion involved in cell communication	0	2	17	↑	<0.001
GO:0002934	desmosome organization	0	2	19	↑	<0.001
GO:0050769	positive regulation of neurogenesis	2	9	1664	↑	<0.001
GO:0050773	regulation of dendrite development	1	6	698	↑	<0.001
GO:0090399	replicative senescence	0	2	21	↑	<0.001
GO:0010517	regulation of phospholipase activity	0	4	258	↑	<0.001
GO:0006620	posttranslational protein targeting to membrane	0	2	24	↑	<0.001
GO:0043517	positive regulation of DNA damage response, signal transduction by p53 class mediator	0	2	26	↑	0.001
GO:0031344	regulation of cell projection organization	3	11	2687	↑	0.001
GO:0098911	regulation of ventricular cardiac muscle cell action potential	0	2	29	↑	0.001
GO:0035082	axoneme assembly	0	3	134	↑	0.001
GO:0036159	inner dynein arm assembly	0	2	34	↑	0.001
GO:1903963	arachidonate transport	0	2	35	↑	0.001
GO:2000192	negative regulation of fatty acid transport	0	2	36	↑	0.001
GO:0003341	cilium movement	0	3	152	↑	0.001
GO:0009968	negative regulation of signal transduction	5	13	3839	↑	0.001
GO:0045597	positive regulation of cell differentiation	4	12	3347	↑	0.001
GO:0090398	cellular senescence	0	3	158	↑	0.001
GO:0006921	cellular component disassembly involved in execution phase of apoptosis	0	3	161	↑	0.001
GO:0006789	bilirubin conjugation	0	1	1	↑	0.001
GO:2000193	positive regulation of fatty acid transport	0	2	43	↑	0.001
GO:0060192	negative regulation of lipase activity	0	2	44	↑	0.001
GO:0032303	regulation of icosanoid secretion	0	2	44	↑	0.001
GO:0030520	intracellular estrogen receptor signaling pathway	0	3	180	↑	0.002
GO:0034350	regulation of glial cell apoptotic process	0	2	49	↑	0.002
GO:0061304	retinal blood vessel morphogenesis	0	2	54	↑	0.002
GO:0032091	negative regulation of protein binding	0	3	201	↑	0.002
GO:0010769	regulation of cell morphogenesis involved in differentiation	2	8	1838	↑	0.002
GO:0007275	multicellular organismal development	26	39	20612	↑	0.002
GO:1900207	negative regulation of pronephric nephron tubule development	0	1	2	↑	0.003
GO:0072013	glomus development	0	1	2	↑	0.003
GO:0072158	proximal tubule morphogenesis	0	1	2	↑	0.003
GO:2001030	negative regulation of cellular glucuronidation	0	1	2	↑	0.003
GO:0039004	specification of pronephric proximal tubule identity	0	1	2	↑	0.003
GO:0043586	tongue development	0	2	63	↑	0.003
GO:0001539	cilium or flagellum-dependent cell motility	0	2	65	↑	0.003
GO:0006182	cGMP biosynthetic process	0	2	68	↑	0.003
GO:0030198	extracellular matrix organization	1	6	1176	↑	0.004
GO:0007399	nervous system development	13	23	10176	↑	0.004
GO:0070980	biphenyl catabolic process	0	1	3	↑	0.004
GO:0048729	tissue morphogenesis	4	10	2907	↑	0.004
GO:0010876	lipid localization	2	6	1176	↑	0.004
GO:0044802	single-organism membrane organization	4	11	3414	↑	0.004
GO:1901571	fatty acid derivative transport	0	2	77	↑	0.004
GO:0097037	heme export	0	1	4	↑	0.005
GO:0071711	basement membrane organization	0	2	83	↑	0.005
GO:0042770	signal transduction in response to DNA damage	0	3	275	↑	0.005
GO:0061298	retina vasculature development in camera-type eye	0	2	86	↑	0.006
GO:0061333	renal tubule morphogenesis	0	3	285	↑	0.006
GO:0086091	regulation of heart rate by cardiac conduction	0	2	91	↑	0.006
GO:0051240	positive regulation of multicellular organismal process	7	14	5233	↑	0.006
GO:0072302	negative regulation of metanephric glomerular mesangial cell proliferation	0	1	5	↑	0.006
GO:0072166	posterior mesonephric tubule development	0	1	5	↑	0.006
GO:0009972	cytidine deamination	0	1	5	↑	0.006
GO:2001076	positive regulation of metanephric ureteric bud development	0	1	5	↑	0.006
GO:0046087	cytidine metabolic process	0	1	5	↑	0.006
GO:0007356	thorax and anterior abdomen determination	0	1	5	↑	0.006
GO:0051552	flavone metabolic process	0	1	5	↑	0.006

GO:0098901	regulation of cardiac muscle cell action potential	0	2	96	↑	0.007
GO:0048858	cell projection morphogenesis	6	13	4751	↑	0.007
GO:0003338	metanephros morphogenesis	0	2	100	↑	0.007
GO:0086002	cardiac muscle cell action potential involved in contraction	0	2	100	↑	0.007
GO:0045664	regulation of neuron differentiation	3	9	2689	↑	0.007
GO:0072112	glomerular visceral epithelial cell differentiation	0	2	101	↑	0.008
GO:0071307	cellular response to vitamin K	0	1	6	↑	0.008
GO:1900142	negative regulation of oligodendrocyte apoptotic process	0	1	6	↑	0.008
GO:0032962	positive regulation of inositol trisphosphate biosynthetic process	0	1	6	↑	0.008
GO:2000533	negative regulation of renal albumin absorption	0	1	6	↑	0.008
GO:0097241	hematopoietic stem cell migration to bone marrow	0	1	6	↑	0.008
GO:0035776	pronephric proximal tubule development	0	1	6	↑	0.008
GO:0035802	adrenal cortex formation	0	1	6	↑	0.008
GO:0061098	positive regulation of protein tyrosine kinase activity	0	2	102	↑	0.008
GO:0072010	glomerular epithelium development	0	2	103	↑	0.008
GO:0051387	negative regulation of neurotrophin TRK receptor signaling pathway	0	1	7	↑	0.009
GO:1902771	positive regulation of choline O-acetyltransferase activity	0	1	7	↑	0.009
GO:1902997	negative regulation of neurofibrillary tangle assembly	0	1	7	↑	0.009
GO:1902960	negative regulation of aspartic-type endopeptidase activity involved in amyloid precursor protein catabolic process	0	1	7	↑	0.009
GO:1902955	positive regulation of early endosome to recycling endosome transport	0	1	7	↑	0.009
GO:1902948	negative regulation of tau-protein kinase activity	0	1	7	↑	0.009
GO:0052697	xenobiotic glucuronidation	48	57	41891	↑	0.009
GO:0044763	single-organism cellular process	0	3	340	↑	0.010
GO:0032368	regulation of lipid transport	2	6	1416	↑	0.010
GO:0043010	camera-type eye development	0	2	117	↑	0.010
GO:0002089	lens morphogenesis in camera-type eye	1	6	2927	↑	<0.001
GO:0002474	antigen processing and presentation of peptide antigen via MHC class I	0	3	342	↓	<0.001
GO:0019882	antigen processing and presentation	0	3	651	↓	<0.001
GO:0006958	complement activation, classical pathway	0	2	342	↓	0.002
GO:0002757	immune response-activating signal transduction	0	3	1532	↓	0.002
GO:0038096	Fc-gamma receptor signaling pathway involved in phagocytosis	0	2	394	↓	0.002
GO:0072376	protein activation cascade	0	2	430	↓	0.002
GO:0002440	production of molecular mediator of immune response	0	2	458	↓	0.003
GO:0016064	immunoglobulin mediated immune response	0	2	513	↓	0.004
GO:0042742	defense response to bacterium	0	2	550	↓	0.004
GO:0034165	positive regulation of toll-like receptor 9 signaling pathway	0	1	25	↓	0.004
GO:0050778	positive regulation of immune response	0	3	2038	↓	0.005
GO:0006959	humoral immune response	0	2	620	↓	0.005
GO:0002682	regulation of immune system process	1	4	4405	↓	0.005
GO:0002449	lymphocyte mediated immunity	0	2	795	↓	0.008
GO:0051707	response to other organism	0	3	2536	↓	0.009
GO:0002460	adaptive immune response based on somatic recombination of immune receptors built from immunoglobulin superfamily domains	0	2	830	↓	0.009
GO:0048251	elastic fiber assembly	0	1	52	↓	0.009
GO:0038095	Fc-epsilon receptor signaling pathway	0	2	844	↓	0.009
GO:0009607	response to biotic stimulus	0	3	2602	↓	0.009

Terms that were also enriched in the spleen are shaded in gray.

Chapter II

Supplementary data (read statistics for each sample, full lists of DEG, and overrepresented GO terms) is available from the Dryad Repository: _________________________________

Table SIII.1 Sampling date, size parameters, and parasitological parameters for all sampled eels.

No.	Date	Body mass (g)	Total length (cm)	Swim bladder: *Anguillicola crassus* adults	larvae (L_3+L_4)	sum	capsules	Gut *Bathriocephalus claviceps*	*Proteocephalus macrocephalus*	Cestodes	*Camallanus lacustris*	Gills *Pseudodactylogyrus* spp.	*Myxidium giardi*	Fins *Ergasilus gibbus*	*Myxobolus portucalensis*	Eyes *Diplostomum* sp.
1	08.08.2017	91	38.5		1	1	4					++	+		++	
2	08.08.2017	255	51.5	6	6	12					2	+		2		
3	08.08.2017	282	55.0	12		12		2		2		+++	++		+++	1
4	08.08.2017	80	37.0	2	2	4						++				
5	08.08.2017	212	49.5	2	3	5	5					+++		1	+++	1
6	08.08.2017	253	55.5		1	1	7	3		3		+++			++	
7	08.08.2017	138	44.0		21	21	3	1		1		++	+	1	+	
8	08.08.2017	56	32.5		5	5						+	+		+	
9	08.08.2017	134	44.5		4	4	3					++	++			
10	08.08.2017	259	56.0	1	1	2						++	+	1	+	
11	08.08.2017	67	36.0	3	3	6	2	1		1		+	+	1	++	
12	08.08.2017	202	50.0		20	20	1?					++			+++	
13	08.08.2017	98	39.0									+	+			
14	10.10.2017	62	34.0	3		3						++		1		
15	10.10.2017	41	30.0		3	3						+	+			
16	10.10.2017	117	40.0	2		2	1					+		2	+	
17	10.10.2017	105	41.5		5	5	2	1	1	2		+	+	2		
18	10.10.2017	83	38.0	11	35	46		2		2		+	+			
19	10.10.2017	75	36.0		3	3						+	+			
20	10.10.2017	122	40.0		7	7	4					+	+	1		
21	10.10.2017	110	40.5		10	10	1		1	1	1	+	++	3		
22	10.10.2017	70	36.5	1	5	6						+	+	2		
23	10.10.2017	42	30.0	2	8	10						+	+			
24	10.10.2017	151	44.5				8	1		1	1	++	+	1		
25	10.10.2017	80	36.0									+	++			
26	17.10.2017	58	34.5		2	2	7	1		1		+	++			
27	17.10.2017	45	31.0	2	2	4	3					+	+			
28	17.10.2017	161	47.0	1	4	5	1	2		2		++	+			
29	17.10.2017	95	41.0	1	6	7		1		1		++	+			
30	17.10.2017	33	28.5		3	3						+	++			
31	17.10.2017	175	46.5	1	4	5						++	+++		++	1
32	17.10.2017	363	57.5	25	7	32						+				
33	17.10.2017	166	47.5	3	12	15						+	+			
34	17.10.2017	58	33.0	3	15	18						+	++			
35	17.10.2017	51	32.5	3	1	4		1		1		+	+			
36	17.10.2017	58	33.5	3		3						+	+			
37	17.10.2017	133	43.0				1	1		1		+	+++	1	+	
38	17.10.2017	124	42.5	4	6	10	1?					+	+			1

Abundances of *Pseudodactylogyrus* spp., *M. giardi*, and *M. portucalensis* were categorized: 1-5 individuals = +, 6-20 = ++, > 20 = +++; ? indicates uncertain encapsulation status, cestodes is the added number of *B. claviceps* and *P. macrocephalus*; individuals 12, 13, 24, 25, 37, and 38 were excluded from analyses, because living *A. crassus* were not present in the swim bladder of because of uncertain encapsulation status